AN ILLUSTRATED HISTORY OF
FIRE ENGINES

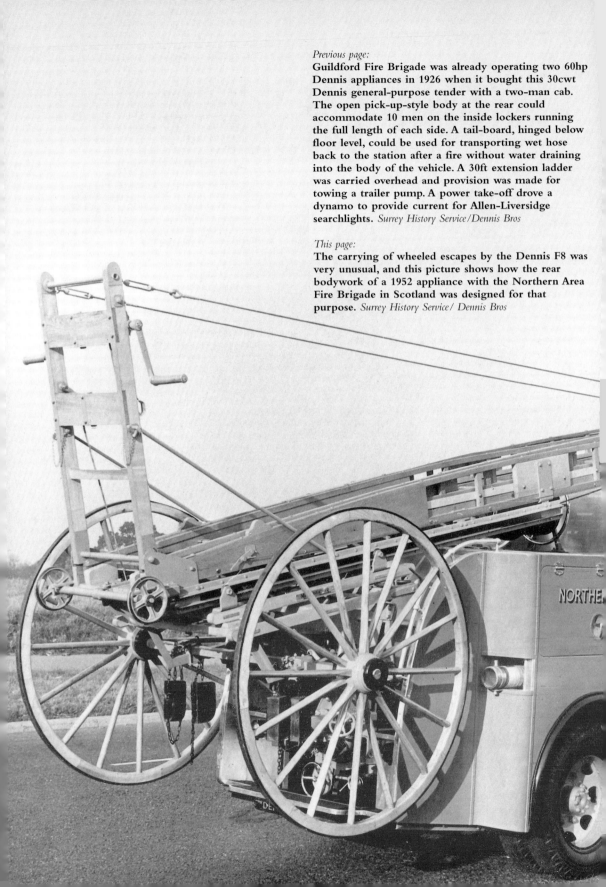

Previous page:
Guildford Fire Brigade was already operating two 60hp
Dennis appliances in 1926 when it bought this 30cwt
Dennis general-purpose tender with a two-man cab.
The open pick-up-style body at the rear could
accommodate 10 men on the inside lockers running
the full length of each side. A tail-board, hinged below
floor level, could be used for transporting wet hose
back to the station after a fire without water draining
into the body of the vehicle. A 30ft extension ladder
was carried overhead and provision was made for
towing a trailer pump. A power take-off drove a
dynamo to provide current for Allen-Liversidge
searchlights. *Surrey History Service/Dennis Bros*

This page:
The carrying of wheeled escapes by the Dennis F8 was
very unusual, and this picture shows how the rear
bodywork of a 1952 appliance with the Northern Area
Fire Brigade in Scotland was designed for that
purpose. *Surrey History Service/ Dennis Bros*

AN ILLUSTRATED HISTORY OF
FIRE ENGINES

ROGER C. MARDON

Ian Allan
PUBLISHING

CONTENTS

Front cover:
This Dennis F48 water tender ladder was new to Suffolk & Ipswich Fire Service in 1969 and sold to the Felixstowe Dock & Railway Company in 1983.
Roger Pennington

Back cover, top:
1939 Leyland FK8 Braidwood pump delivered to Budleigh Salterton UDC Fire Brigade and now preserved. *Roger C. Mardon*

Back cover, bottom:
1984 Dodge G13/Saxon water tender ladder of Avon Country Fire Brigade. *Roger C. Mardon*

First published 2001

ISBN 0 7110 2483 9

Published by Ian Allan Publishing

an imprint of Ian Allan Publishing Ltd, Hersham, Surrey KT12 4RG.

Printed by Ian Allan Printing Ltd, Hersham, Surrey KT12 4RG.

Code: 0111/B

ACKNOWLEDGEMENTS

I would like to thank the manufacturers and fire brigades who have helped and co-operated in the preparation of this book, together with all members of the Fire Brigade Society, the Fire Service Preservation Group and others who have in any way contributed to its completion.

For their expert knowledge and unhindered access to historic material, special thanks are due to: Tom White and John Meakins of the Kent Fire Brigade Museum; Steven Shirley and Brian Harris of the Manston Fire Museum; John Arthur, formerly of the Royal Air Force Fire Service; Malc Tovey formerly of Leicestershire Fire & Rescue Service; Roy Goodey and Phil Stevens, past and present librarians of the Fire Service Preservation Group; Stuart Brandrick, former director of Carmichael Fire; Aidan Fisher, custodian of the HCB-Angus archives; Laurence Spring and Sally Jenkinson of the Surrey History Service, custodian of the Dennis archives; and Dennis Shearer of the Vauxhall Heritage Centre.

The historic Leyland photographs supplied for the book are the copyright of the British Commercial Vehicle Museum Trust Archive.

BIBLIOGRAPHY

Fire and Water magazine, 1892-1913

Fire magazine, 1908-2001 (currently published by dmg world media [uk], Redhill)

Manual of Firemanship (London: HM Stationery Office, first published 1944)

Barclay, T. Denis, ed Roy Goodey, *Fire Service Pioneers* (London: The Watchroom, 1999)

Blackstone, G. V., CBE, GM, and Bentley, Ronald C., *A History of the British Fire Service* (Borehamwood: The Fire Protection Association, Jubilee edition 1996)

Burgess-Wise, David, *Fire Engines and Fire-fighting* (London: Octopus Books, 1977)

Gamble, Sidney Gompertz, FSI, AMICE, MIFireE, *A Practical Treatise on Outbreaks of Fire* (London: C. Griffin & Co Ltd, 1926)

Georgano, G. N., *The Complete Encyclopedia of Commercial Vehicles* (Iola, Wisconsin: Krause Publications, 1979)

Gilbert, K. R., MA, DIC, *Fire Fighting Appliances* (London: HM Stationery Office, 1969)

Guttenberg, Paul, *The History of the Turntable Ladder* (Ulm, Germany: Magirus-Deutz, undated)

Holloway, Sally, *Courage High! A history of firefighting in London* (London: HM Stationery Office, 1992)

Jackson, W. Eric, *London's Fire Brigades* (London: Longmans, 1966)

Jenkinson, Keith A., *Preserved Lorries* (London: Ian Allan, 1977)

Kenyon, James W., *The Fourth Arm* (London: George G. Harrap & Co, 1948)

Merryweather & Sons Ltd, various illustrated catalogues

Merryweather, James Compton, MIME, *The Fire Brigade Handbook* (London: Merritt & Hatcher, 1888)

Morris, Major C. C. B., CBE, MC, MIMechE, *Fire!* (London: Blackie & Son, 1939)

Shaw, Capt Eyre M., *Fire Protection* (London: Charles and Edwin Layton, 1890)

Woodcroft, Bennet, *Alphabetical Index of Patentees of Inventions* (London: Evelyn, Adams & Mackay, 1969 [first published 1854])

Young, Charles F. T., CE, *Fires, Fire Engines and Fire Brigades* (London: Lockwood & Co, 1866)

GLOSSARY

AFFF	aqueous film-forming foam
AFS	Auxiliary Fire Service
ARP	air-raid precautions
BA	breathing apparatus
BAA	British Airports Authority
BCF	bromochlorodifluoromethane, a vaporising liquid used as a fire extinguishing medium
bhp	brake horse power
CBM	chlorobromomethane, a vaporising liquid used as a fire extinguishing medium
cc	cubic centimetres
CO_2	carbon dioxide
CTC	carbon tetrachloride, a vaporising liquid used as a fire extinguishing medium
cwt	hundredweight
ft	foot/feet
gpm	gallons per minute
gvw	gross vehicle weight
hp	horse power
IFE/CFOA	Institution of Fire Engineers/Chief Fire Officers Association
in	inch(es)
JCDD	Joint Committee on Design and Development
lb	pound(s) (weight)
l/min	litres per minute
LPG	liquefied petroleum gas
m	metre(s)
methyl bromide	a vaporising liquid used as a fire extinguishing medium
MIRA	Motor Industry Research Association
mm	millimetre(s)
MoD	Ministry of Defence
mpg	miles per gallon
NFS	National Fire Service
ohv	overhead valve
psi	pounds per square inch
PTO	power take-off
RDC	Rural District Council
RIV	rapid intervention vehicle
UDC	Urban District Council

The convention employed when describing fire appliances is for the vehicle classes to be mentioned first, followed by the bodybuilder/supplier and then the manufacturer of any special equipment or fire engineering. Imperial or metric units of measurement have been quoted generally according to the manufacturers specifications of the time.

PREFACE

This history of the fire engines of the British Isles includes those built elsewhere provided they were in service here, with necessary reference to some foreign inventions and developments for the sake of completeness. I make no apology to readers who may decry use of the term 'fire engine', for it is a phrase that everyone understands, and 'engins' (sic) for the extinguishing of fires were constructed long before the first 'fire appliance' rolled off the production line. Apart from a few machines that pre-date photography, the illustrations used are contemporary with the subject and all vehicles were in service when photographed. The history would be incomplete without the inclusion of a number of landmark appliances, but I have tried to use as many previously unpublished photographs as possible.

While I believe that my research has been thorough, it is worth pointing out that original sources do not always agree. So, for example, when a manufacturer's documents state categorically that the first of a particular type of appliance was built for Leicester in 1913, and Birmingham is equally positive that it had one two years earlier, there is obviously room for manoeuvre, and some dates may therefore be a year or two out. I would be pleased to know of any material or photographs that may confirm or deny the accuracy of my account or assist with any future publication.

Roger C. Mardon
Canterbury, Kent
June 2001

INTRODUCTION

To understand fire engines and the way they work it is first necessary to understand fire. For fire to be sustained three things must be present: fuel, oxygen and heat. Take away one of these and the fire will go out. For example, fire can be starved of fuel by pulling apart a burning thatched roof or by creating fire-breaks in a forest fire. It can be starved of oxygen by smothering it to keep out fresh air, as in snuffing out a candle or applying a layer of foam to burning oil. It can be starved of heat through cooling of the burning material by the application of a cooling substance such as water. Fire engines provide the medium for fire-fighters to do these things, and a good many more besides.

The Romans had initiated organised fire-fighting of a sort by 300BC, and by the 1st century AD had established the Corps of Vigiles to protect their city. Their equipment may have been basic, but it was based on the same understanding of fire that we have today. Long hooks and axes were used to pull down buildings in the path of the fire to stop it spreading, and bucket chains were organised to throw water on the fire and to supply the simple pumps or siphons employed by the Vigiles.

It is inconceivable that the Romans would not have brought their knowledge and organisation to Britain during their 400-year occupation of these lands, but when they left early in the 5th century it is clear that their skills left with them. The danger of fire, however, remained only too apparent, and as the centuries passed fire prevention measures were introduced to regulate the hazardous practices of residents and traders alike and to control building custom. Fire hooks, axes and buckets remained in use, but any form of pump, or organised fire-fighting, had become a thing of the past.

From the 16th century some municipal authorities made requirements for the provision of fire-fighting equipment, but it was the Great Fire of London in 1666 that aroused the nation's interest in fire protection. Even then, most people were more concerned about avoiding personal financial loss, and so began fire insurance.

While the London authorities remained reluctant to establish any system of organised fire-fighting, the insurance pioneers soon saw the folly of standing by while buildings burned, and set up their own private fire brigades. The existence of different insurance brigades, each with their own interests at heart, led to rivalry. This was not conducive to efficient fire-fighting, which required co-ordination and a spirit of co-operation, and it led to the amalgamation of brigades under the London Fire Engine Establishment.

The primary concern of the insurance brigades was the saving of property to reduce the level of claims. The saving of life from fire was very much a secondary concern, so in 1836 the Royal Society for the Protection of Life from Fire was established in London, absorbing the earlier Fire Escape Society and taking over its wheeled escape ladders. The Metropolitan Fire Brigade was formed in 1866 and maintained out of public funds, and this took over responsibility for escapes in London, by which time the number of escape stations had reached 85.

Although local authorities were empowered to provide a fire brigade, many failed to do so, or to do so with any efficiency, and private brigades continued to protect commercial premises and country houses. Such brigades often rendered services beyond their gates as a means of gaining fire-fighting experience or out of the proprietor's desire to help the community. Though declining in numbers, works and occupational fire brigades continue to this day, particularly at concerns posing special risks such as oil refineries and chemical plants.

This state of affairs continued until the 1930s when, encouraged by the threat of war, the Government introduced legislation for the compulsory provision of fire brigades by local authorities. In England and Wales 1,440 fire authorities were established by the Fire Brigades Act 1938, with another 228 in Scotland. The prospect of war also prompted the authorisation of an Auxiliary Fire Service in 1938 to augment the local authority brigades. World War 2 and experience of the Blitz brought forth calls for nationalisation of the fire service, and emergency legislation to that end was approved by Parliament in 1941. The National Fire Service was thus born on 18 August 1941, with 33 fire areas across England and Wales and six in Scotland. The AFS was absorbed into the National Fire Service.

The Government kept its promise to return the fire service to local authority control after the war, but avoided re-establishing the multitude of brigades that had previously existed. The new fire authorities were to be centred on the 146 counties and county boroughs of England and Wales, joint schemes reducing the number of brigades to 135, which became operational on 1 April 1948. In Scotland 11 new brigades became effective on 16 May. Under separate legislation, Northern Ireland had four fire authorities, but in 1950, apart from Belfast which reverted to a municipal brigade, the province became protected by the Northern Ireland Fire Authority. The appliances of the NFS were reallocated to the new brigades and more than a little discontent was caused when some expensive machines funded at the ratepayers' expense were not returned to their towns of origin.

The AFS, as part of the National Fire Service, disappeared when fire brigades were returned to local authority control but, as part of the Government's response to the threat of nuclear war, the Civil Defence and Auxiliary Fire Service were promptly reconstituted under the Home Office. Recruitment started in 1949 with the idea that both would be made up of civilian volunteers. Over the years, as the threat receded, the AFS was seen as an unnecessary expense and it was disbanded in 1968. Some of its appliances were sold off, some were mothballed in Home Office depots, and a relative few were bought by local authority brigades.

Since then there has been a succession of local government reorganisations leading to the 62 fire brigades now serving the United Kingdom, the Channel Islands and the Isle of Man.

Below:
This 1999 Dennis Sabre/GB Fire/Magirus turntable ladder is described more fully on page 82. *Dennis Fire*

1. EARLY DEVELOPMENTS

Manual Pumps

Until recently water has always been regarded as plentiful and cheap, so putting it on a fire has long been recognised as an effective and inexpensive way of extinguishing the fire. It can be thrown on with a bucket or applied in greater force and quantity with a pump. After the departure of the Romans, it is apparently not until the 16th century that the use of squirts in this country can again be demonstrated. These came in different sizes and capacities from half a gallon to maybe 3 gallons, and were made of brass. It took one or two men to hold and direct the device, and a third to pull or push the piston handle.

In Europe the first recorded wheeled fire engine was built for the German town of Augsburg by goldsmith Anton Plater in 1518. A wheeled syringe engine is described in Besson's *Théâtre des Instruments* authorised in 1568, although it is commonly known as Lucar's engine because of the description and picture in *A Treatise Named Lucarsolace* devised by Cyprian Lucar, Gentleman, published in England in 1590. It comprised a cone-shaped container mounted on a two-wheeled carriage, which could be filled with water through a funnel on top. The water was forced out of a pipe at the end of the cone by turning a screw handle, which pushed forward a piston within the container. There is evidence to show that large squirts on wheels were in use in England after that.

After squirts and screw-operated force pumps came lever-operated pumps. These were worked manually by alternately raising up and pushing down on a lever to operate a piston in the pump barrel. Simple force pumps mounted on a sled or on wheels had been around since early in the 17th century, as recorded by Saloman de Caus in 1615 and John Bate in 1634. With each stroke of the pump piston these engines produced a spurt of water that could be directed on to the fire through a swivelling pipe fixed to the top of the apparatus. In the most common design, the pump mechanism was housed in a cistern, which was filled by bucket chain. Such pumps were operated by no more than four or five men working a single pump handle up and down, with force enough, it was said, to project the water up to 40ft in the air.

Similar engines appear to have been in use in various parts of the country during the first half of the 17th century, and some would have been available in London at the time of the Great Fire in 1666, along with squirts, buckets, hooks and ladders.

In 1672 the Dutch engineer Jan van der Heiden invented a new type of fire engine that could draw water from an open source through a suction hose and discharge it through a leather delivery hose, which could be taken in close to the fire. Several of these new engines were brought to England when William and Mary arrived from the Netherlands to claim the

Below:
In the 1670s John Keeling of Blackfriars was building engines in the form of a wheeled-tub or cistern containing a pump worked by two levers and with a central swivelling delivery pipe or gooseneck. One such was supplied for use in Dunstable in 1678 and remained in service until 1841.

throne in 1688 and more were imported until at least 1790.

More force could be exerted by more men if two levers were linked and worked by men on opposite sides of the pump. As one lever was depressed so the opposing lever was raised, and with a rhythmic rocking motion new designs meant that an efficient stream or jet of water could be produced. It became generally accepted that 30 or 40 strokes a minute was a realistic working rate, but it was hard work and the men could work for only about 5 minutes at a time. The pumping teams were then changed over, and while one team pumped the other would be refreshed with supplies of beer.

It is Richard Newsham who is best remembered in Britain for his early manual pumps. He put the levers at the sides of the machine rather than across the ends as had previously been the norm, allowing more men to work. Another innovation, which did not last, was the use of treadles so that in addition to the men working the levers others could stand on top of the frame working the treadles up and down with their feet. Newsham manufactured six sizes of engine, ranging from a 30-gallon cistern with a discharge capacity of 30 gallons per minute up to a 170-gallon cistern with a capacity of 170gpm worked by 20 or more men.

While a number of fire engine patents were registered in the 18th century, Newsham's direct competitor was Fowke of Wapping, who also seems to have produced six sizes of engine. The larger sizes could produce two streams of water at once, which Fowke advanced as a major selling point.

From the end of the 17th century Nathaniel Hadley built fire engines in Cross Street, London, and moved to a new factory at Long Acre in 1738. The business was later carried on by his son, also Nathaniel, to be joined by master plumber Charles Simpkin, together trading as Hadley & Simpkin, and then by Henry Lott, when the firm became Hadley, Simpkin & Lott. Moses Merryweather was taken on as an apprentice in 1807, marrying Lott's niece in 1836 and taking over the firm to trade under his own name after Hadley, Simpkin and Lott had all retired. His first son, Richard, was apprenticed in 1855 and went on to play a prominent part in the company. After Moses died in 1872, followed by Richard only five years later, his second son, James Compton, succeeded as head of the business, which was incorporated as a private limited company in 1892. Merryweather & Sons Ltd continued for nearly another century as a well-known name in the fire engineering business.

Adam Nuttall started a fire engine works in Long Acre in about 1750 and both he and Samuel Phillips built in the same style for a time at least. Phillips was succeeded by Hopwood in 1798. W. J. Tilley took over from Hopwood in 1820 and built engines until 1851, when his business was in turn succeeded by Shand Mason & Co. This company survived until 1922, when it was bought out by Merryweather & Sons.

Joseph Bramah's engine, patented in 1785, departed from the usual operation in that it was a semi-rotary pump. Instead of pistons sliding within a straight cylinder, his design consisted of two wings or blades alternately moving clockwise and anticlockwise within a watertight case. With a system of valves, water was drawn in and forced out by the alternate semi-rotary action of the wings. Roundtree was another builder using the semi-rotary pump after the style of Bramah in the last quarter of the 18th century.

Left:
This Newsham manual was built in 1734 with two single-acting 4⅓in pumps and an air vessel, a device to even out the flow of water from the pump to produce a continuous stream rather than spurts. Instead of a swivelling branch, the leather hose and branchpipe could be connected to a delivery outlet on top of the tall casing enclosing the air vessel.
Science Museum/Science & Society Picture Library

The sleds or small wheels of the early pumps allowed a degree of manoeuvrability at the fire, but did not provide an efficient carriage for hand-hauling the machines to the fire. Thus it was common to put the engine on a larger hand or horse-drawn cart for carrying to the fire, where it would be off-loaded before getting to work. It seems to be around the end of the 18th century that pivoted axles enabling engines to be steered came into use, and, at about the same time, Hadley & Simpkin fitted road springs and hinged the pump handles, or levers, so that they folded over at each end to avoid the overhang that made it difficult to manoeuvre the engines. Until the beginning of the 19th century the use of horse-drawn fire engines was not widespread, but with these improvements, and as engines got larger, it was a natural progression for them to be built as horse-drawn vehicles.

Merryweather produced the London brigade engine for the Great Exhibition of 1851, and in 1866 was making five sizes of horse-drawn brigade engines, so called because they were deemed suitable for use by public fire brigades. The largest was said to deliver 220gpm to a height of 150ft when worked by 46 men, while the smallest, known as the Paxton, was worked by 22 men and delivered 100gpm to a height of 120ft. Merryweather's country brigade engine had a smaller hose box and suction pockets than the London brigade model, but still provided seating for the men. Hand-hauled village engines mounted on springs, mansion engines without springs, and factory engines without hose box or suction pockets were each available in nine sizes from six-man to 46-man operation. The Greenwich manual, incorporating various technical improvements, was introduced in 1890.

In 1866 Shand Mason was building five sizes of London brigade engine to be drawn by one or more horses for town use, three country brigade engines to be pulled by one horse, five hand-drawn factory engines from 16-man to 46-man operation, and five hand-drawn mansion engines requiring from eight to 30 men.

In the middle of the 19th century it was popular to test different makes and models of engine against each other. The results were of dubious value because it was impossible to guarantee the equality of strength and effort of opposing teams, who were, in any event, working at maximum output for short periods to win a competition rather than in a sustained fashion to extinguish a fire. Rates of 60 strokes a minute or more were recorded, but it would have been impossible to keep that up for long.

The advent of steam power, and even motor power, did not herald the immediate end of the manual pump. The last fire in London at which a manual pump was used was in 1899, but four years later Shand Mason was still advertising its 22-man brigade manual with seating for eight to 10 men. In May 1923 *Fire* magazine reported that some fire brigades had taken the pumps out of manuals and fitted them on motor chassis, using the motor for both tractive and pumping purposes. But Boston Fire Brigade in Lincolnshire left its pump as manually operated after mounting it on a Ford motor chassis, and was proud of the anachronism.

Steam Pumps

It was not until 1829 that John Braithwaite and John Ericsson of London built the first steam fire engine. It was mounted on a four-wheeled carriage for horse traction and weighed 45cwt complete, about twice the weight of an equivalent manual. It could raise steam from cold in 13 minutes and would pump 150gpm to a height of 90ft, but the London Fire Engine Establishment steadfastly refused to adopt the principle of steam power. Braithwaite's second steamer was exported, but in 1831 Liverpool had the third, which remained in use for many years.

Shand Mason built the first British land steamer after Braithwaite in 1858, but this was exported to Russia. The London Fire Engine Establishment used its first steamer in 1860, a Shand Mason weighing over 4 tons that required three horses to pull it. It was replaced within 12 months. In the same year James Shekleton of Dundalk produced Ireland's first steamer.

Merryweather's first land steamers, the 'Deluge' in 1861 and the 'Torrent' in 1862, both went to the Hodges' Distillery Fire Brigade at Lambeth. In 1862 Shand Mason built two steamers for the London Fire Engine Establishment, and William Roberts of Millwall built his first two steam fire engines. One was the self-propelled machine mentioned later and the other,

'Princess of Wales', was taken into use by the London Fire Engine Establishment. Merryweather introduced the 'Sutherland' in 1863, a horizontal double-cylinder engine that was bought by the Admiralty for use in Devonport Dockyard, where it remained in service until 1905. Shand Mason introduced the vertical engine in the same year.

After Liverpool, the first town outside London to acquire a steamer was Alton in Hampshire, which bought a 250gpm single-cylinder Merryweather in 1864. That year Shand Mason built two steamers for London and one for Dublin. The early steamers did not deliver vast quantities of water and something like 200gpm at 100lb per square inch pressure proved to be realistic and popular. The competitive spirit of the manufacturers led to bigger engines with greater capacity, but fire brigades generally settled on something up to 500gpm.

Shand Mason introduced the 'Equilibrium' pattern in 1869, an engine of three steam cylinders and three double-acting pumps arranged vertically, and the first was supplied to Glasgow in 1870. Edinburgh got its first steamer in 1873, but reputedly the funnel was too tall for the machine to be kept in the engine house and it was kept in a shed half a mile away, to be used only in dire emergency. By the time Charles Young had written his invaluable work *Fires, Fire Engines and Fire Brigades* in 1866, Shand Mason & Co had established itself as market leader in the manufacture of steam fire engines. The company had built 60 of all kinds, in comparison with Merryweather's 17 land-based and three floating engines. Merryweather introduced the 'Metropolitan' single-cylinder for London in 1884 and the 'Greenwich' double-cylinder in 1885, the first of which went to Manchester the following year. In 1889 Shand Mason introduced the double-vertical engine, which survived as the standard configuration until the end of steam power.

Merryweather and Shand Mason had doubtless secured the market and only a few other smaller manufacturers ventured forth, but not for long. William Rose & Co of Manchester constructed its first steam fire engines in 1897, 350gpm machines delivered to Selly Oak and King's Norton. The company went on to make 250gpm and 450gpm models, but discontinued fire engine building in 1902.

Steamers were not always coal-fired. In 1900, after a series of experiments, the Metropolitan Fire Brigade decided to change to oil-firing, and by 1902 had nine oil-burning engines and a further six on order. The engines carried from 2½ to 3 hours supply of oil. A conversion kit with a new burner using steam from the boiler to spray oil into the furnace, ready for fitting to a horsed engine, cost from £35 up to £140 according to horsepower.

Often the fire engine would be kept on a gentle ramp in the fire station to help the horses overcome the

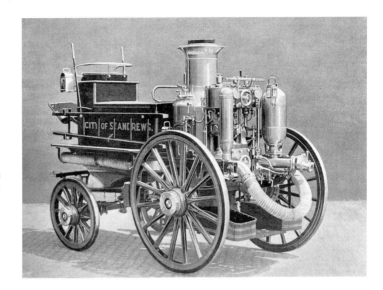

Right:

The Merryweather 'Greenwich Gem' steamer, introduced in 1896, was available in four sizes from 225gpm to 500gpm. The machinery was placed vertically behind the boiler, with a stoke hole at the back of the boiler so that the fire could be tended while the engine was travelling. The engine was supplied with four copper branchpipes and up to six nozzles, lamps, water bags for the wheels, and various tools and fire irons, all for between £400 and £550 according to size. *Author's collection*

initial inertia when moving off, and even today an appliance ready and able to answer a call is said to be 'on the run'.

In 1905 Shand Mason & Co was offering steamers with apparatus for quick steam raising, which rendered stoking en route unnecessary. The fire would be lit on leaving the fire station and the engineer standing on the rear footboard would operate a pulley wheel and handle fitted at the back of the steam cylinders to revolve an exhauster fan in the funnel at high speed. This, it was claimed, reduced the time to raise 100lb steam pressure from cold by 2 minutes, to 5 or 5½ minutes. Merryweather's rapid steam-raising attachment comprised a cylinder of compressed carbon dioxide gas bracketed to the engine near the boiler and used to increase the draught given to the fire. The manufacturers claimed advantage over other systems because no hand labour was necessary and no

mechanical attachments had to be removed after use, suggesting that Shand Mason's fan did not remain in situ once steam had been raised.

London's last horse-drawn steamer, a Merryweather, was delivered in 1902, but steamers were not finished. At the end of 1912 nearly 450 Shand Mason double-vertical steamers were in use, and as late as 1923 some brigades were still ordering horse-drawn steamers when it was not unusual for chief officers to prefer a steamer for use in rural areas. It was not until 1932 that the National Fire Brigades Association dropped steamer drills from its annual competitions, although a number of steamers remained in service into World War 2.

During World War 1 the requisitioning of horses by the Army left many fire brigades unable to call on their usual supplier. This led to manual and steam pumps being towed behind motor vehicles, something for which they had never been designed, and accidents

Right:

Shand Mason's quick-steaming device is seen on this No 2 size double-vertical engine of 450-500gpm capacity, which was supplied to York Fire Brigade in 1907. *'Fire & Water' magazine, by courtesy of Kent Fire Brigade Museum*

occurred as a result. The heavy steamer, often still fitted with its iron tyres, was prone to skid, jack-knife or even overturn unless extreme care was exercised while towing, especially when going downhill or round corners.

Perhaps the most significant development of the steam fire engine came in 1841 with the completion of the first self-propelled steamer, built in New York by English engineer Paul R. Hodge. This was also the first steam fire engine constructed in the United States.

In 1901 and 1902 C.T. Crowden of Leamington Spa converted a number of horse-drawn steamers to self-propulsion by the addition of a propelling double-cylinder engine to provide power for chain drive to the

Left:
Probably the best-known self-propelled steamer was Merryweather's 'Fire King', first built in 1899 for Port Louis, Mauritius. Leyland, Lancashire, was the first town in Britain to buy one in 1903, and within four years another 21 were on the run in Great Britain. A coal-burning furnace was an option, but the boiler was usually oil-fired with two fuel tanks fitted under the driver's seat. The vehicle was propelled by chain-driven rear wheels and could travel at 20 to 30mph, and its 'Gem' pattern pump was available in six sizes ranging from 300gpm to 1,000gpm. Three of Liverpool's appliances and the Edinburgh one were also provided with a 60-gallon chemical engine and 180ft hose-reel. *Author's collection*

Left:
In 1902 the Metropolitan Fire Brigade workshops adapted a Merryweather steamer for self-propulsion. This served at Whitefriars and was used with some success in inner London on level roads, but it was unable to negotiate a significant gradient without stopping at the bottom to build up a sufficient head of steam. It was converted back to horse traction in 1906. *London Fire Brigade*

Right:
Shand Mason did not produce its first self-propelled steamer until 1909. An oil-burning boiler provided steam to a double-cylinder reversing engine, which powered the live back axle through a chain drive to propel the vehicle at up to 20mph. An independent double-cylinder variable expansion pattern 500gpm fire pump, with four delivery outlets, was located in front of the propelling engine. The first was delivered to London and the second went to Bristol. *London Fire Brigade*

rear wheels. The Norwich Union insurance brigade in Worcester is known to have had one.

A year after its first effort, the London brigade replaced the fore-carriage of a horse-drawn steamer with a four-wheeled petrol-driven tractor, upon which the front end of the steamer was mounted to create an articulated vehicle. However, it had a tendency to jack-knife, and was sold in 1909. London Fire Brigade acquired its first Merryweather 'Fire King' in 1905 at a cost of £1,150. This was a 500gpm machine weighing about 5¾ tons and running on solid rubber tyres. With 80 gallons of reserve water for the boiler, it could travel 9 miles without taking on water, consumed 2 gallons of paraffin per mile, and could pump for 2 hours without refuelling. By 1908 London had six self-propelled steamers, the later ones being capable of 400gpm.

Steam power was combined with the internal combustion engine in 1913 when Shand Mason demonstrated a 400gpm double-vertical steam pump mounted on a modified 40hp Daimler car chassis. This appliance had accommodation for men and equipment, including a telescopic ladder, and was fitted with a 40-gallon chemical engine. A number of steam pumps were installed on motor chassis by brigades wishing to take advantage of the petrol engine without the expense of buying a whole new fire engine.

Hose Carts and Hose-reels

As piped water supplies improved and hydrants became more common in the mid-19th century, some town fire brigades, having the benefit of reasonably constant water pressure and not offering protection to neighbouring rural areas, were able to operate by connecting their hoses directly to a standpipe set into a hydrant. This avoided the expense of buying a manual or steam engine, and instead they would have a hose cart or a hose-reel cart to provide stowage for the hose and accessories.

The hose cart was more suited to towns with closely spaced hydrants, where small runs of hose were the norm. The simplest was a box, usually with a hinged lid, mounted on a two-wheeled hand-drawn carriage and able to carry 300-800ft of flaked hose, a standpipe, branchpipe, nozzles and hand tools. It might even have carried a set of scaling ladders and been fitted with lamps. A larger quantity of hose could be carried by a one-horse hose cart, comprising a two-wheeled vehicle with shafts for the horse, a driver's seat and room for another man to sit or stand on the back.

The hose-reel cart, an American idea with coupled hose kept wound on a reel, was preferred where longer runs of hose had to be laid. The reel, usually of wood,

Right:
This Merryweather 'Newcastle' pattern hose cart supplied to Wimbledon UDC incorporates a box with a hinged cover at the front for tools and small gear, and a well at the rear in which hose could be carried either coiled or flaked. Such a model would have cost about £18 without equipment.
Author's collection

was mounted between the two wheels of a carriage and carried between 100 and 1,000ft of leather hose, or between 300 and 2,000ft of canvas hose. An equipment box would normally be mounted across the top in line with the axle, but narrow versions were available with the box placed lengthways to negotiate passages less than 3ft wide. Hand hose-reels were more common, but the larger sizes were available for horse traction.

Where a steam pump was arranged to carry only the driver, the stoker and working tools, a hose tender was also needed to transport men, hose and equipment to the fire. Four-wheeled vehicles having the appearance of horse-drawn manual pumps were built as hose tenders but, as no pump was installed, the whole of the internal space was available for hose and equipment. Such a tender might carry 1,500ft of hose in the main box, a jumping sheet, hand pumps and small gear in a front compartment, and standpipes, hydrant keys and long items in side pockets or lockers. A rear well might hold other equipment or the coal required to keep the steamers going. Combined hose and coal tenders were common.

Ladders and Escapes

Ladders and escapes have long been associated with fire-fighting and rescue from fire. Manual and steam pumps often carried scaling ladders, about 6ft 6in long and narrower at the top, enabling several to be slotted together for greater length. Hook ladders, or pompier ladders, originated in France and became common in England during the 1880s, although they were not adopted in London until 1903. There were different designs but essentially they comprised a ladder about 13ft 6in long with a barbed hook at the top that held the ladder on a window-sill. A fireman could thus scale the face of a building by hooking the ladder over a first-floor sill, climbing up to sit astride the sill, raising the ladder to the next sill, and so on until he reached his goal; their use was not discontinued until the 1980s. E. H. Bayley & Co supplied London with its first telescopic extension ladders in 1890, and ladders of various lengths were carried on hand and horse-drawn carts. Early ladders were manufactured from timber, but alloy rapidly increased in popularity after World War 2, and of course extension ladders still feature strongly in today's fire service. Modern fire engines also carry one or two alloy roof ladders.

The wheeled escape had been introduced as a life-saving appliance reaching to the second floor in 1818, run to the fire by a conductor and helpers on its own carriage wheels some 5ft in diameter. By 1836 the carriage was sprung, a 19ft fly ladder, hinged near the top of the main ladder, could be raised to give more height, and a canvas rescue chute, or shoot as it was called, was often provided for people to slide down. A supplemental ladder could be attached to give a working height of about 51ft. The fly ladder escape remained common until 1880 when the extending escape came into general use. In 1848 the Royal Society for the Protection of Life from Fire was operating a wheeled escape with a cage or balcony that could be winched up and down the ladder. The ladder was a single section only, but a first-floor ladder carried with the escape could be attached to the cage enabling the third floor of a building to be reached. Sliding carriage escapes, enabling different angles of elevation to be achieved, were introduced in London during 1893, and the capital's last fly ladder was used in 1897. Horse-drawn escape carriages appeared from 1890. These became known as escape vans, possibly because they were at the front, or vanguard, of the turn-out rather than from the nature of the vehicle, which was nothing like an enclosed van. Within two or three years it was quite common for escapes to be carried on hose carts.

Horse-drawn manual pumps were adapted for the carriage of escapes and, indeed, were offered from new as combination engines. To make room for the ladders

Left:
Shand Mason built a new free-standing, 40ft, light, hand-hauled four-wheel escape for Alton and this 50ft version for Cromer in 1908. The ladder comprised four sections of the company's lattice girder design, with a plumbing arrangement. It could be elevated and extended over the length of the carriage at a steep angle for directing jets on to the fire, or at a lower angle for bridging forecourts and other obstacles, but there was no turntable enabling rotation.
'Fire & Water' magazine, by courtesy of Kent Fire Brigade Museum

in the centre, the positions of the coachman's and officer's seats were extended sideways to overhang the front wheels of the cart. In 1907 a new 26-man Merryweather 'Greenwich' manual pump was available with a choice of seven escapes from 35 to 65ft, the mid-range 50ft model costing £300.

London's last horsed escape was withdrawn in 1920, but an escape was hand-run to its final fire in 1922.

Chemical and First Turn-out Engines

However prompt the brigade was in turning out and reaching a fire, there would be some delay before water was applied while the fire engine was either set into open water or connected to street hydrants and the hoses coupled. Towards the end of the 19th century the chemical engine was introduced to provide an instant means of attacking a fire. This generally comprised a four-wheel horse-drawn carriage similar to a manual engine. Indeed, they were often converted from manual pumps, but two-wheel versions for horse or hand drawing were also available. One or two cylinders containing some 40 or 60 gallons of water, into which bicarbonate of soda had been dissolved, were mounted on the carriage. Sulphuric acid was released from a bottle within the cylinder and the contents mixed by turning a built-in agitator; the chemical reaction of the acid on the carbonate solution produced carbon dioxide, which pressurised the cylinder just like a soda-acid portable fire extinguisher. The water would be forced out for 8 to 15 minutes by the gas pressure and was directed on to the fire by a hose. This might be sufficient in itself to extinguish the fire but, if not, it allowed time for more substantial apparatus to be got to work.

The benefit of combining a chemical engine with an escape was spotted almost immediately, and machines were developed under a variety of different names — first turn-out engines, escape tenders and hose-reel escapes, to quote a few. In 1898 Shand Mason & Co introduced its first alarm machine, which comprised a four-wheeled horse-drawn carriage with accommodation for six firemen and the driver. This used

compressed air to pressurise the water cylinders, with a shut-off valve incorporated in the air supply enabling the water jet to be interrupted or shut down altogether. This was an improvement on the soda-acid system, which continued to discharge until the cylinders were empty.

At the turn of the century steam power was in its heyday and mains water supplies were more widespread. As a result many town brigades had little use for the perfectly serviceable manual engines they still had, except perhaps as hose tenders. For one or two hundred pounds they could have a manual pump converted into a chemical engine, and in 1902 Merryweather installed a 40-gallon chemical engine and 120ft hose-reel in such a pump for Bristol Fire Brigade.

Early in the 20th century first-aid appliances were being built on petrol-engined chassis. Eccles in Lancashire is credited with having bought the first motor vehicle for fire brigade use, having ordered it in April and taken delivery in September 1901. It was a first-aid tender based on the Bijou car chassis built and supplied at a cost of £180 by the Eccles-based Protector Lamp & Lighting Company. In 1902 C. T. Crowden was building a Daimler-chassised petrol-engined tender for his home town of Leamington Spa. This appliance, reported to have been delivered in March 1903, accommodated a crew of six and was equipped with a chemical engine, hose and a 30ft wheeled escape.

In June 1903 Merryweather supplied Tottenham District Council with a petrol-driven combination appliance, which is often said to have been the first of its kind. However, it seems clear that Crowden's machine for Leamington Spa was delivered three months earlier. In July the Metropolitan Fire Brigade followed suit with a 10-12hp Wolseley, which initially

Right:
In 1904 Shand Mason & Co introduced a new first turn-out machine. This was a combined escape and four-wheel hose carriage with a 40-gallon chemical engine able to provide a strong jet for 10 minutes. The example shown was designed for quick turn-out and rapid conveyance by a pair of horses, but the engine was available in various sizes for draught by one horse or by a pair of horses. *'Fire & Water' magazine, by courtesy of Kent Fire Brigade Museum*

Left:
**Wolseley Motors provided
Leicester with its first motor
appliance, a chemical first turn-
out engine built in 1904. The
vehicle weighed 3 tons and
could make up to 24mph. This
£650 machine stayed on the run
until 1913.**
Malc Tovey

had a small pump driven from the road engine. This was
soon removed, however, and the vehicle was used
successfully as a first-aid tender. Also in 1903 John
Morris & Sons of Salford produced its first motor
tender.

In 1905 Merryweather constructed a petrol-engined
escape and compressed air fire engine for the London
Fire Brigade, and the next year London adopted
mechanical traction for all new appliances. The first
were 50hp first turn-out machines, or motor escape
vans as London preferred to call them, carrying a 55ft
wheeled escape and equipped with a 60-gallon
chemical engine, 180ft hose-reel and other hose and
hydrant gear.

Electrically propelled fire engines were popular
elsewhere in the world, where they could be plugged
into the tramway electricity supply in the street or
powered by batteries. In Britain they were in limited
use by 1911, but were generally considered to be
impractical. The batteries were heavy and took up space
on the vehicle, which anyway was restricted in the
distance it could travel on one charge, particularly in
hilly areas.

Merryweather had produced the first motorcycle
combination first-aid appliance in 1910, and
production of such appliances seems to have continued
until the early 1930s, finding more favour abroad than
at home. In 1921 the Evinrude Motor Co of London
was offering a Coventry-Victor motorcycle and sidecar
first-aid combination. In 1924 Birmingham firm
Dunford & Elliott Ltd announced a motorcycle
combination first-aid engine, the Dunelt-XLCR.
Equipment included a 25-gallon chemical engine with
25ft of hose, three 3-gallon 'Sprafoam' extinguishers, a

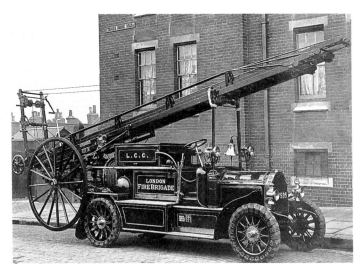

Left:
**One of London's early escape
vans was this Belsize of 1906,
equipped for fire-fighting by
John Morris & Sons and
sporting the solid rubber tyres
with steel-studded fabric inserts
that were popular at the time.
However, it seems to have been
a one-off and the brigade sold it
to Southampton in 1916 for
£300.** *Surrey History Service/
Dennis Bros*

2-pint CTC extinguisher, and a collapsible 12ft ladder. About the same time, Leyland Motors secured a demountable self-contained motor pump in the sidecar position of a BSA motorcycle combination to create a first-aid appliance. Suction, several lengths of delivery hose, branches and nozzles were carried, and a transverse seat for a second crew man was also fitted. In 1932 Dennis Bros adapted a Matchless motorcycle combination to produce a first-aid appliance carrying a portable pump, suction, delivery hose and nozzles.

Right:
The principle of powering a vehicle by electric motors fitted to the front wheels was developed by Ferdinand Porsche, adopted by his employer Jacob Lohner of Vienna, and used for fire engines early in the 20th century. In 1906 the designs were sold to Emile Jellinek, who hired Porsche and went on to produce Mercedes Electrique battery-electric vehicles at Austro-Daimler. Cedes Electric Traction marketed the chassis in England until about 1916, and a number of chemical escape vans were built for the London Fire Brigade. In 1919 the brigade had 11 electric escape vans in service. *Eric Billingham collection / London Fire Brigade*

Right:
This 1912 Dennis motor first-aid tender for Gosforth Fire Brigade clearly shows the chemical engine installed behind the seat, with the hose-reel above. Also clearly visible is the pedal-operated gong operated by the officer. *Surrey History Service / Dennis Bros*

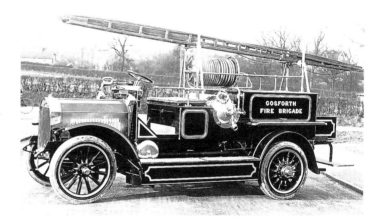

Right:
This two-seater motorised tricycle was a product of Simonis in 1915. At least one portable extinguisher is attached to the front and a ladder is secured to the side by buckled leather straps. *Fire Service Preservation Group*

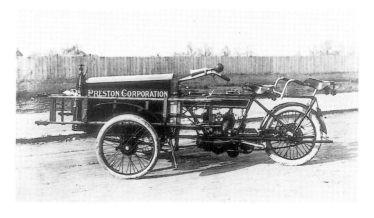

With the advent of the motor-driven auxiliary pump able to supply a hose-reel from an on-board water tank, the chemical engine began to disappear. It had given good service for 10 or 12 years, but few were built after about 1910, and many brigades converted their first-aid tenders into full pumps to give their appliances both first-aid and main pumping ability. Brigades that still had a need for a first-aid appliance turned to tenders built on a light commercial chassis incorporating an auxiliary pump, a first-aid tank, generally up to 60 gallons, and a hose-reel. Such vehicles were often produced by local firms, such as St Anne's Automobile Co of St Anne's-on-Sea, which produced the Cotton-Hall tender, and Martin Walter Ltd of Folkestone, which constructed light tenders on 1-ton chassis. Accommodation for crew and equipment was provided and provision was often made for towing a trailer pump.

Two examples serve to illustrate other appliances that fell short of being motor pumps. In 1930 Wolstanton

Fire Department commissioned a special service and first-aid van built by Simonis Ltd on a Ford chassis. Fitted with a first-aid pump and 100-gallon tank and a Foamite foam generator, this appliance was presented as particularly suitable for work in rural areas where, it seems, bus and coach fires were on the increase. The following year, Congleton in Cheshire acquired an unusual forward-control Thornycroft escape carrier with a New World body by the Lawton Motor Body Building Company of Stoke-on-Trent. Access from the back being impossible owing to the escape and Dennis trailer pump carried on board at the rear, the crew entered via a platform behind the nearside front wheel.

The escape-carrying unit was a wartime emergency appliance built on the Austin K4 or Fordson 7V chassis to carry a wheeled escape. It was equipped with a 130-gallon water tank and twin 120ft hose-reels, and provided with a special tow-bar for drawing a trailer pump. The tow-bar had to extend behind the carriage and wheels of the escape, and the original design was

Below:
Touring cars were often adapted for fire brigade use, and Count Louis Zborowski, a car and motor-racing enthusiast who was killed in the 1924 Italian Grand Prix at Monza, donated this Rolland-Pilain to his home village of Bridge, Kent, in the early 1920s. Coachbuilders Blythe Brothers, of Canterbury, adapted the body to

accommodate 10 men and hose, and the vehicle was used to tow a manual pump that had been converted into a trailer; whether this was a modified version of the brigade's 19th-century Merryweather is unknown. The car was later replaced by a Rolls-Royce donated by another local benefactor. *Kent Fire Brigade Museum*

Right:

This reconditioned Austin towing vehicle and 'Hatfield' trailer pump were supplied by Merryweather to Saxmundham Fire Brigade in 1929 and represent the light motor car conversions of the day. *Nicholas Blake and Keith Francis collection*

Left:

This hose-reel tender was commissioned by Acton Fire Brigade in 1934, on the unusual forward-control version of the Dennis Ace. *Surrey History Service/Dennis Bros*

Below:

John Kerr & Co (Manchester) supplied this Ford V8 tender, with bodywork by J. H. Jennings & Son of Sandbach, and 250–500gpm Drysdale trailer pump to Lymm UDC Fire Brigade, Cheshire, in 1936. *Stuart Brandrick collection*

prone to bending as the appliance went round sharp corners. This problem was overcome, but did not arise at all on later appliances, which were fitted with a Barton front-mounted pump.

After the end of World War 2 many auxiliary towing vehicles were converted into hose-reel tenders by the installation of a 120-gallon water tank, hose-reel pump and 180ft hose-reel fed out through portholes in the appliance body. In 1949 there were 1,038 towing vehicles in service in England and Wales, most of which had been modified in this way. New appliances intended for towing trailer pumps in rural and semi-rural areas were covered by specifications drawn up by the Home Office Joint Committee on Design and Development . They needed a first-aid tank of between 100 and 150 gallons, one hose-reel and a 25gpm pump to supply it. A self-contained light portable pump was optional, but accommodation for a crew of six was required. These appliances were commonly called hose-reel tenders.

As late as 1963 Bristol Fire Brigade found a need for a new hose-reel tender, and this was built by HCB on a Bedford J 3-ton chassis. A 30gpm hose-reel pump was supplied by a 150-gallon water tank, but a Coventry Climax portable pump, on a nearside slide-out rack, could be operated from the appliance.

Left:
World War 2 generated a need not only for the production of fire pumps in great numbers but also, in the case of trailer pumps, for just as many towing vehicles. In the early stages of the conflict taxis and other light vehicles were used and, with the experience gained from such use, the auxiliary towing vehicle was developed. This was built on a 2-ton chassis with a steel body providing seating inside for the crew and locker space for equipment. The roof was specially strengthened to afford protection from shrapnel and other flying debris. This atmospheric picture shows a column of Austin K2 auxiliary towing vehicles and their pumps under escort by a despatch rider.
The Westbrook collection

Left:
This Morris Commercial hose-reel tender was one of two bodied by Wadham and bought by Berkshire & Reading Fire Brigade in 1954. *Charles Keevil*

2. MOTOR PUMPS

Prewar Pumps

The first petrol-engined pump was built by Daimler of Germany in 1888 and was in fact a horse-drawn appliance after the style of a steamer, but with an internal combustion engine at the rear instead of a boiler and steam engine. The first motor pump delivered for use in Britain was an 1895 Daimler four-wheel trailer pump, which went to the Datchet estate of the Hon Evelyn Ellis.

It was inevitable that the advent of the petrol-engined motor vehicle would see a change in fire engine design. Motorised first turn-out machines and escapes had proved successful and motor pumps soon followed.

The pump itself was either a reciprocating pump or a centrifugal pump (often described as a turbine pump). Reciprocating pumps were the same as those fitted to steamers, but now driven by a petrol motor rather than a steam engine. The centrifugal pump, of which Merryweather claimed to have built the first for a South American city in 1906, consists of an impeller with vanes rotated at speed within a casing. As water is introduced into the casing from the centre it is propelled outwards by centrifugal force and discharged at pressure through the outlet. More than one can be coupled together to form a multi-stage pump, which is more efficient than one large single-stage pump. The centrifugal pump remains the most widely used type of fire-fighting pump to this day.

Johann Zwicky, a Swiss engineer, built a huge machine for Tottenham Fire Brigade with the aid of the council workshops in 1906. Believed to have been built on a chassis by Simms Manufacturing Co Ltd of Kilburn, the fire engine is reputed to have eventually cost £3,000, equating today with a price of £176,000. A Simms four-cylinder engine supplied the motive power. Two 500gpm pumps coupled together are said to have provided two deliveries each side, and the 1¼in nozzle on the rotating gear-elevated monitor could project a jet 100ft high. The pumps were apparently unable to take water from hydrants, presumably therefore relying on open water. The machine had a 60-80-gallon chemical engine and hose-reel.

In 1908 John Morris of Salford produced its first turbine motor pump, a Belsize-chassised machine that went to Bury in Lancashire. Confirming the move away from the chemical engine, this was fitted with a small auxiliary pump to supply a hose-reel from the appliance's built-in water tank.

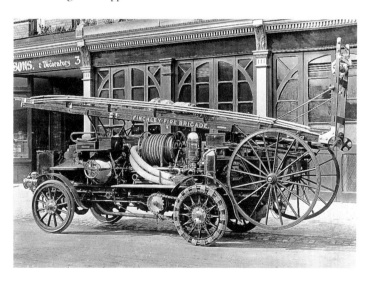

Right:
Having built the world's first fire engine with a petrol engine providing power for both propulsion and operation of the fire pump for Baron de Rothschild's French estate in 1904, later that same year Merryweather supplied Finchley Fire Brigade in north London with a similar machine. Capable of over 20mph, this had a four-cylinder 24/30hp engine with chain drive to the rear wheels through a cone clutch and gear. The combination apparatus, as Merryweather called it, carried a wheeled escape and was fitted with a 'Hatfield' 250gpm three-cylinder reciprocating pump, a 60-gallon soda–acid chemical engine and 180ft hose-reel, and a hose box. The appliance underwent a number of alterations in its life and is seen here in a rare view after replacement of the original engine but before steering modifications and the addition of twin rear wheels. A similar machine was adopted by Aberdare in South Wales.
Surrey History Service/Dennis Bros

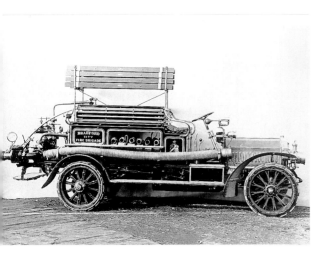

Left:
John Dennis built a bicycle during his spare time in 1893 and with brother Raymond soon established a successful business. In 1898 they made their first motor-tricycle, three years later their first car, and three years after that their first commercial vehicle. The first fire engine built by Dennis Bros of Guildford was supplied to Bradford in 1908 at a cost of £900 and remained in service until 1922. It was powered by a White & Poppe engine and utilised a rear-mounted Gwynne centrifugal pump. A set of four scaling ladders was carried on gantries over the hose and equipment box. *Surrey History Service/Dennis Bros*

Left:
Starting as the Lancashire Steam Motor Co in 1896, Leyland Motors Ltd was incorporated in 1907 and went on to produce petrol-engined and steam wagons side by side. The company's first fire engine, which had a rear-mounted Mather & Platt centrifugal pump but carried no ladder, was delivered to Dublin Fire Department in 1909. It was based on the larger of two four-cylinder petrol-engined chassis produced by the new company, and this U-type chassis remained the basis of Leyland fire engines until the FE range appeared in the early 1920s. *The British Commercial Vehicle Museum Archive*

Left:
In 1909 Harrow took delivery of a 400gpm Merryweather pump, much the same as that adopted by London the year before. In 1929 this similar machine, very likely the same one, was refurbished by Merryweather. Early Merryweathers were built with imported Aster running gear by Ateliers de Construction Mécanique l'Aster of St Denis, France. *Nicholas Blake and Keith Francis collection*

Right:
The first motor pump built by
Halley Industrial Motors Ltd of
Glasgow, a six-cylinder vehicle
with a 450gpm Mather & Platt
pump, was delivered to Glasgow
Fire Brigade in 1910. The second
machine went to Leith later in
the year, and the example seen
here was bought by Aberdeen
Fire Brigade in 1912.
Roy Goodey collection

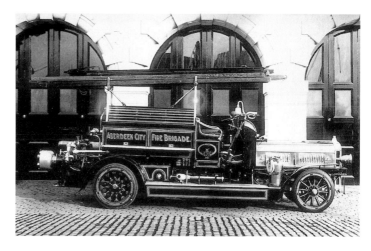

In 1911 London Fire Brigade commissioned its first centrifugal pump from Dennis Bros. A four-cylinder 40hp engine powered the machine and the rear-mounted Gwynne 350gpm pump was driven through a power take-off. The Argyll Motor Co Ltd of Alexandria, Scotland, built the first of its only two motor pumps in 1911 and supplied it to Dundee the year after. J. & E. Hall Ltd, with origins going back to the Dartford Iron Works of 1785, provided one of its Hallford chassis for an escape carrier of the Sidcup Fire Brigade in 1912. The appliance, which was bodied by Beadles of Dartford, remained in service until 1936.

Right:
Vehicles with six-cylinder
engines were characterised by
long bonnets, and this 1913
Belsize is no exception.
Described at the time as a
motor turbine pump combined
hose tender and ladder carriage,
this machine was fitted with a
600gpm pump and supplied by
John Morris & Sons at a cost of
£1,150. It was scrapped in 1939,
but not before it had been given
a new Dorman engine in 1928.
Malc Tovey

Right:
Nottingham also opted for the
greater power of six cylinders,
choosing Dennis for this pump
escape of 1912. The escape is a
Shand Mason lattice girder
pattern. *Surrey History Service/
Dennis Bros*

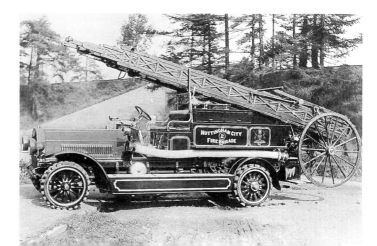

Tilling-Stevens of Maidstone was a proponent of the petrol-electric chassis, where the petrol engine, rather than driving the vehicle direct, powered an electric dynamo, or generator, the power from which was used to provide the drive. By 1919 the company had designed a fire engine where the dynamo powered the road wheels or supplied current to an electro-turbine pump. The novel feature of this appliance was the demountable 350gpm pump, fitted with large carriage wheels, which could be wheeled by hand to a water source, or wherever, and powered by electric current from the motor dynamo through a cable.

After World War 1 there was a demand for light motor pumps suitable for smaller towns and country districts, which were often unable to afford a new motor fire engine. Indeed, if they could readily secure the use of horses, many were happy to continue using a reliable steamer. Others looked to the conversion of a motor car or lorry chassis for brigade use, and Sinclair & Co of east London was one firm undertaking the necessary work. Many surplus military vehicles had come on to the market, and in 1920 the company converted a 30-35hp Fiat army lorry into a motor tender for Friern Barnet Council. Meanwhile, J. & E. Hall of Dartford was offering a fire tender body, complete with chemical tank and 50ft wheeled escape, to suit a 30cwt chassis.

Dennis Bros brought the first Tamini pumps to England from Italy in 1921, and was soon building them in Guildford, opening the way for a range of British and foreign cars to be adapted as fire engines. By 1924 Brighton had two Fiats with Tamini pumps, while

Left:
Founded by O. Simonis in 1903, Simonis of London built fire engines on a drop-frame Commer chassis produced by Commercial Cars Ltd. This picture, taken at East Ham fire station, probably in about 1916, shows a Commer/Simonis pump and pump escape. *Roy Goodey collection*

Below:
A Leyland chassis with a four-cylinder engine powered this pump built for Chorley Fire Brigade in 1919. Some six years earlier Leyland Motors had opened a new plant at Chorley for fire engine production and it was appropriate that the local brigade should be so equipped. *The British Commercial Vehicle Museum Archive*

a 200-250gpm Hatfield pump had been fitted to a Lorraine-Dietrich touring car for Swindon Fire Brigade by Merryweather in 1922. Martin Walter Ltd of Folkestone was building 200gpm light motor pumps and John Morris & Sons introduced a little Guy-chassised appliance with an 'Ajax' 150-250gpm pump mounted amidships. Tilling-Stevens, too, found the Guy convenient for its light pump.

In order to provide a vehicle of greater load capacity, a standard Ford chassis could be lengthened to 13ft 6in by the addition of a rear extension frame from Baico Patents Ltd. The existing back axle was fitted with a sprocket for a chain drive to the Baico rear wheels mounted on the extension. The extended chassis was able to carry a load of 30cwt and was therefore suitable for carrying a light escape. A pump

and first-aid tank, with auxiliary pump and hose-reel, could also be fitted.

The first motor pumps were provided with solid tyres, but by 1922 pneumatic tyres were in use on small motor pumps such as the Dennis 250/300gpm turbine motor pump. Merryweather, too, had fitted pneumatic tyres in 1922, but heavier fire engines were generally fitted with solid tyres that were perceived to offer capability at speed without excessive fuel consumption. However, in the mid-1920s pneumatic tyres were gaining in popularity for heavy commercial vehicles, and manufacturers such as Dunlop began to promote them for fire service use. In 1928 came London's first fire engine with pneumatic tyres, a Dennis 500-600gpm pump, but by then Birmingham, West Ham and Ayr, among others, had already made the transition

Right:
The Stanley Fire Engine Co Ltd of Halifax was offering two light fire engines in 1922 built on Ford chassis. A 1-ton chassis was provided with bodywork to carry a crew of up to eight men with 1,000ft of hose and equipment, as seen here, while a lighter chassis accommodated a crew of four and 500ft of hose. A hand-primed 150gpm turbine pump was centrally mounted on the chassis and driven by a power take-off. *The British Commercial Vehicle Museum Archive*

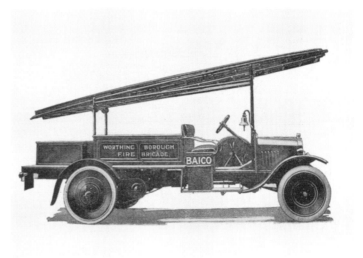

Right:
1922 saw Worthing using a 30cwt Ford with the Baico extended chassis. The advertisement stated, 'One of the most essential requisites for small towns and villages is an efficient Fire Engine at moderate cost, with ample capacity for ladders and gear, and having a speed of 15–35 miles per hour ... carries a really suitable ladder and equipment and is not merely an adaptation of a short, cheap and inadequate chassis.' This model was available with a 200-250gpm front-mounted turbine pump. The Baico–Tonna was announced in 1925. *By courtesy of 'Fire' magazine*

with apparent success. Conversion of pumping
appliances from solids to pneumatics was offered by the
tyre manufacturers, with new wheels and hubs as part
of the package; Twickenham Fire Brigade, for example,
found that its Dennis pump went 4–6mph faster.

A motor pump carrying a wheeled escape was
known as a pump escape. The escape was originally
developed as a rescue ladder and this combination
offered the advantage of making available on one
vehicle the means of extinguishing a fire — the pump
— and the means of saving life — the escape. The
differing attributes of escapes and extension ladders
usually meant that they were not interchangeable, so
early motor appliances were built either as pumps or as
pump escapes. This limitation was later overcome and
dual-purpose appliances, able to carry either an escape
or an extension ladder, became available. Halley offered
such an option in 1923.

Some idea of fire engine use at this time can be
gained by looking at Croydon in 1923, a town covering
some 14sq miles with a population of 200,000. There
was some industry but the principal risks were
associated with the residential nature of the town —
shops, theatres, cinemas, and no fewer than 16 railway
stations in the brigade area. The first motor appliances
were acquired in 1912, a pump escape and a pump. Two
more pump escapes and a tender with light ladder
followed in 1914. Four or five fire calls a week were
answered by the four fire stations, the central station
pump turning out to all fires with a pump escape from
either the central station or a sub-station as appropriate.
The central station appliances covered 400-500 miles a
year. The pump's fuel consumption was about 6mpg
while travelling and 2 gallons per hour while pumping,
giving running and maintenance costs of about £1 a
week. The pump escapes each cost about half this. The
solid tyres on which the machines ran lasted about five
years, after which they were either worn out or
perished.

Leyland's 1923 range of appliances comprised a
200gpm Ford/Stanley pump, for which the company
had that year secured the sole concession, a 30cwt
appliance with a 250gpm Rees-Roturbo pump, the
FE1 36-40hp light appliance with a 300gpm Rees-
Roturbo pump, the FE2 with a four-cylinder 55hp
engine able to pump 500gpm, and the FE3, with a six-
cylinder 90hp engine, had a 1,000gpm capacity.

In industrial areas the fire risks were considerably
greater than elsewhere, with manufacturing processes,
docks and warehousing. Consequently larger fire
engines were generally thought to be most suitable for
these areas and, at the National Fire Brigades
Association conference in 1926, 70hp appliances with
a pumping capacity of 700-1,000gpm were
recommended as the norm. Water supplies were not
usually a problem and metalled surfaces able to
withstand the weight of heavy pumps were
widespread. In other urban areas, made up of office
buildings, shops, hotels and often large older houses, a
500gpm pump was thought to be convenient. A
machine not over about 4 tons would be able to
negotiate traffic without difficulty while retaining the
ability to operate effectively, generally taking water
from hydrants. Suburban residential areas were often
supplied with water from a small 4in main and there
was a greater chance of an appliance having to cross
rough ground. Consequently a lighter engine of only
250gpm capacity was put forward as the ideal, with an
extension ladder in preference to a wheeled escape, as
the latter was difficult to manoeuvre across often wide
front gardens. In rural districts limited money was
available for fire protection, which was frequently

regarded as a low priority anyway. A light 200gpm pump to cover an 8-mile radius was suggested, with a trailer pump or two, these being very useful for off-road working. It was common for more than 1,000ft of hose to be used in rural fires, and pumps able to operate at adequate pressure over a period of several hours were needed.

During the mid-1920s six-wheeled pumps became popular, and Bridport Fire Brigade was among the first to be persuaded that a Morris Commercial six-wheel chassis was 'better for crossing rough ground, to the extent that it would go where a trailer pump could not be manhandled'. The Thornycroft A5 War Office subsidy chassis was another six-wheeler, which Simonis engineered and supplied to Blackburn with a New World body.

The year 1928 was significant for Merryweather because it introduced a centrifugal pump, although it was marketed alongside the company's successful 'Hatfield' reciprocating pump for many years to come.

The design of early motor fire engines reflected the horse-drawn era, and the open style of these appliances was known as the Braidwood body, after James Braidwood of the Edinburgh and London Fire Engine Establishments. A hose and equipment box still formed the bulk of the body behind the front seats, and the crew sat on this while resting their feet on a footboard, or perhaps stood while precariously clinging to the ladder. Not surprisingly firemen were sometimes thrown off as they tried to dress in their fire gear or as machines cornered at speed. Horsed engines were designed to travel at, maybe, 10mph but the motor pump could achieve 50mph or more and at this speed the Braidwood body was completely unsuitable. But in spite of advances in design, open bodywork remained a popular choice for many brigades until World War 2, and as late as 1956 the industrial brigade of Pilkington Glass ordered a Braidwood-bodied Dennis F2 pump.

Right:
This Thornycroft six-wheeled appliance was engineered by Merryweather for the War Office in 1929. Caterpillar tracks, seen stored on the foot locker, could be fitted on the back wheels in a few minutes when abnormal conditions were encountered. In addition to the 'Hatfield' rear-mounted pump, the appliance was supplied with a first-aid hose-reel and ladder.
Nicholas Blake and Keith Francis collection

Right:
The most apparent feature of this 1927 Dennis pump escape of Croydon Fire Brigade is the Simonis escape, which, it was claimed, could be unshipped by one man owing to a swinging frame. It was marketed as a self-supporting water tower escape to meet the needs of those brigades not calling for a single-purpose turntable ladder. At an angle of 78° and fully extended, the ladder supported a test weight of 5cwt and was available with a monitor attachment.
Surrey History Service/Dennis Bros

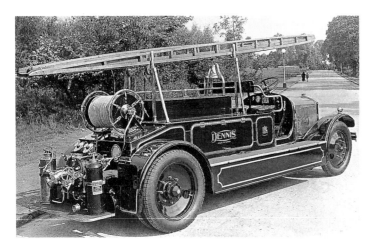

It was not until 1928 that demands for a safer design produced the New World, inside, or wagonette body style, as it was variously called. This was still an open vehicle and the crew sat on inward facing seats down the sides, generally entering from the back. This design enabled the crew to rig in their fire kit en route and removed the risk of being thrown off. Instead of being mounted on a gantry above the equipment box, as it was on a Braidwood-bodied appliance, the ladder would more often be hung on brackets along the outside, where it did not obstruct the crew seating space. The first New World-bodied fire engine in England was a Dennis built for Luton Fire Brigade in 1928.

Right:
Torrington in Devon joined with the neighbouring rural area to buy a Dennis 250gpm light fire engine and a 200-250gpm trailer pump in 1929 — unusually, therefore, the lettering on the appliances showed both UDC and RDC. This skeletal machine was in complete contrast to Birmingham's Dennis New World of the same year and solid tyres were fitted. *Surrey History Service/Dennis Bros*

Transverse seating, where the crew sat on forward-facing seats behind the driver and officer, offered greater safety and comfort for the crew, although they were still exposed to the weather. One of the first manufacturers to incorporate this style must have been Tilling-Stevens, whose light motor tender was in service with Salisbury Fire Brigade and towing a Tilling-Stevens 'Bantam' trailer pump in 1924. London Fire Brigade adopted the transverse style for Dennis escape vans new in 1933.

Morris Commercial moved to the New World design in 1931 with a 250-400gpm pump for Chesham UDC Fire Brigade. The Royal Burgh of Haddington, near Edinburgh, commissioned a Morris

Commercial pump with New World body style the next year. The midships-mounted Gwynne pump had three suction inlets, one each side and one at the front. A 40-gallon tank and first-aid hose-reel were fitted and an extension ladder was carried on the offside.

Right:
The Leyland LTB1 was a coach chassis shortened for fire engine use, and from which was developed the FT range. This 1930 model for Birmingham, with a 50ft escape and first-aid hose-reel, was one of the first and one of the earliest Leylands to be supplied with pneumatic tyres. *The British Commercial Vehicle Museum Archive*

Right:
In 1931 Edinburgh became the first brigade in Britain to put into service an enclosed motor pump. This was built by Merryweather & Sons on an Albion low-load chassis and the saloon body offered the whole crew protection from the elements and the safety of travel without any danger of being flung off as the appliance sped on its way. Seats were provided for 12 men down the sides of the interior with hose lockers beneath, and there was room to stand up inside the saloon, which had folding doors at the back for access. A Merryweather centrifugal pump, capable of delivering up to 800gpm, was installed centrally and an extension ladder and two hook ladders were carried on the roof. *Author's collection*

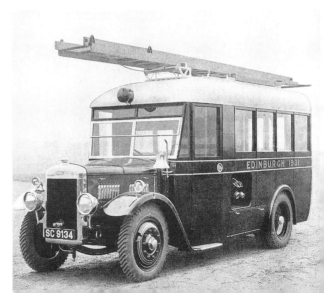

In 1931 Dennis Bros supplied the chassis of a new limousine pump built locally for Darlington Fire Brigade. This was the first saloon-bodied appliance in England and was designed so that all equipment was accessible from the outside. A three-man cab was provided, and the saloon, accessed from a platform doorway at the nearside rear, accommodated 12 crew. The appliance was fitted with a rear-mounted Dennis 300-400gpm pump and, because it was primarily intended for use at fires outside the district, it would tow a trailer pump for water relay. A 30ft extension ladder was carried on the roof. The 16-year-old tender replaced by the new vehicle was converted into a foam and rescue tender, with a 45ft escape for urban work. *Surrey History Service/Dennis Bros*

In 1932 Dundee took delivery of Scotland's first six-wheeled appliance, a Leyland pump with a New World body, which delivered 820gpm at 85psi in tests. Its first fire was at a meal mill outside the brigade district, where it pumped with success from a stream to supply four lines of hose. The following year the appliance responded to a call 14 miles from Dundee in a severe blizzard and over icy roads. After 6 miles the machine ran into a snowdrift and skidded into a ditch. No one was injured or thrown out of the New World body, and within 5 minutes the crew had the vehicle out of the ditch and back on its way.

Bedford was established as a manufacturer of commercial vehicles in 1931 when General Motors created a production line at the Vauxhall Motors factory in Luton. By 1933 the first Bedford fire engines were on offer, built on a 2-ton chassis and equipped with a Pulsometer 350gpm rear-mounted centrifugal pump. A 45-gallon tank and hose-reel formed part of the standard specification and a triple extension 35ft ladder was included. The 350gpm model was priced at £740, but a smaller 250-300gpm was available at £685.

In 1932 Watford Fire Brigade ordered a pump escape to be constructed on a Scammell low-load-line chassis with a 700gpm pump and 55ft Bayley wheeled escape. Delivered the following year, this was to be the first fire pump built by Watford-based Scammell Lorries Ltd. It was powered by a four-cylinder 85bhp engine and came with a Braidwood body. Watford had Scammell's second appliance two years later, but, apart from some wartime ARP vehicles, the company then stayed out of the fire engine business until it took over from Thornycroft as a major supplier of airfield crash tender chassis in 1977.
The British Commercial Vehicle Museum Archive

In 1933 Luton Fire Brigade had two pump escapes built by Vauxhall Motors on Bedford long-wheelbase chassis and driven by a six-cylinder 57bhp engine. The Pulsometer 350-400gpm pump was installed beneath the driver's seat with suction and delivery connections on both sides. A 50-gallon first-aid tank and 250ft hose-reel were fitted, and space for 2,000ft of 2½in canvas hose, breathing apparatus and other gear was provided. The transverse seating style was adopted and a 60ft escape was carried with its wheels brought well forward to counteract backlash. Fittings were chromium-plated and without the escape the appliance cost £850. *Vauxhall Motors*

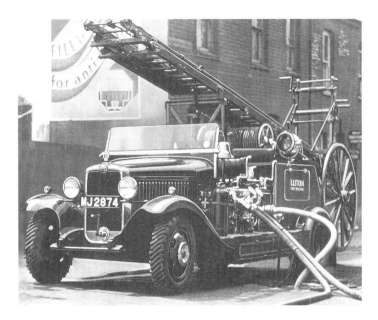

Leyland's standard range of motor pumps in 1933 comprised the six-cylinder 60bhp FK1 with a 400gpm pump, built at Kingston and to be known as the 'Cub', and the low-load chassis 100bhp FT1 with a 700gpm pump. The bigger machine was available in a normal-control bonneted version or in a rarely chosen forward-control version. As the years passed improvements and changes were made, leading to progressive increases in the FK and FT designation numbers. Often these signified whether the pump was rear or midships-mounted — the FK6, for example, had a rear-mounted pump, while that of the otherwise similar FK7 was midships-mounted. The FKT of 1937 was a hybrid combining the 'Cub' chassis with the so-called 'Tiger' engine from the FT range.

Such importance was placed on the ability to start the engine as soon as it was needed that it was common for two or three ignition systems to be provided. In big full-time brigades, the engines would be started up and run for a few minutes every four hours to keep them ready for instant action. By the early 1930s electric engine heaters and battery chargers were available, which kept the engine hot while the appliance stood in the fire station. Some even incorporated a relay to start the engine as soon as the station bells went down.

On a motor pump one engine usually drives both the vehicle and its pump. In road-going mode, power developed by the engine is transmitted via a gearbox to the driving wheels, but when pumping it is diverted to the pump by means of a power take-off and secondary transmission shaft between the gearbox and the pump. While rear-mounting of the pump provides easy access

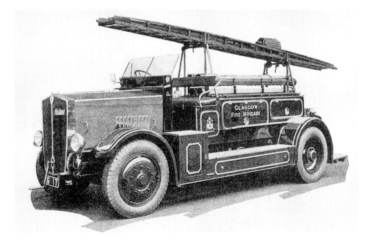

In 1933 Halley Motors delivered two motor pumps to Glasgow Fire Brigade, bringing the total number of Halley appliances there to 11. The pump itself was a rear-mounted Drysdale of 550-600gpm capacity.

By courtesy of 'Fire' magazine

for pump maintenance, it was not convenient on an appliance carrying a wheeled escape because the escape had to be removed from the vehicle — or slipped, as a fire-fighter would say — to give the pump operator unhindered access to his pump. In the 1930s, therefore, it became common for escape-carrying appliances to have the pump installed between the front and back wheels with its controls at the side of the vehicle, just as the cylinders of chemical escapes had been in the days of horse-drawn machines. A direct drive could be employed and midships-mounted pumps were satisfactory from an operational point of view, although accessibility to the pump mechanism for maintenance was reduced.

Front-mounted pumps found lasting favour in America and mainland Europe, but apart from the touring car and light commercial conversions in the 1920s, were unusual in Britain until World War 2. The advantage of a front-mounted pump is the ability to drive it directly from the crankshaft of the road engine, avoiding the complication of a power take-off, but it was always seen as being more vulnerable to accidental damage. To increase their versatility many utility escape-carrying units of World War 2 were fitted with the American-made Barton pump, which was installed

at the front. After the war it was common practice for Merryweather 60ft turntable ladders to be fitted with a similar pump, and in 1953 Hampshire Fire Service decided to install Barton pumps on 12 new water tenders it planned to build.

In 1933 London Fire Brigade broke with tradition and ordered four Dennis and two Merryweather pump escapes, the brigade having relied up to then on motor pumps and separate escape vans with hose-reel equipment. This followed the experience of a 60-pump fire at Rum Quay, resulting in six other fires having to be dealt with by escape vans only. However, established practice was not yet completely forsaken, as two Dennis motor pumps and four Merryweather and four Dennis escape vans were also ordered.

In 1934 London's first limousine pump was ordered from Dennis Bros, the appliance to have a rear-mounted pump and the connected suction carried in a tunnel on the nearside. Flaked delivery hose was carried in a locker above the pump. At this time London had just decided to carry breathing apparatus on machines other than rescue tenders, and the Dennis limousine would provide accommodation for three sets.

The first Dennis Ace fire appliance was a New World model that left the factory for Ascot in 1934, although

Left:
This Dennis pump escape was delivered to London in 1934. It provided transverse seating and the pump was rear-mounted, with the suction pre-connected in the London style.
Surrey History Service/Dennis Bros

Left:
Open transverse seating was rare on Leyland's smaller fire engines, but this FK1 'Cub' supplied to Scarborough Fire Brigade in 1934 features such an arrangement with a midships-mounted pump. The ladder carried is a Morris 'Ajax'.
The British Commercial Vehicle Museum Archive

the chassis had been introduced in the previous year as a bus. The short bonnet and front wheels set well back, which contributed to the compactness and manoeuvrability of the machine, earned it the nickname of 'Flying Pig'. It was powered by a four-cylinder 3,770cc engine and was priced between £500 and £1,000 depending on specification. The model survived until 1939 and the 134th and last appliance built went to Arundel in Sussex.

In 1935 Dennis Bros introduced the Ace 6 chassis, popularly known as the Light 6. This was powered by a six-cylinder 7,413cc engine and lasted until 1943, by which time 84 had been built at prices from £790 to £1275, according to specification. The Leyland FK6, with a new six-cylinder ohv engine and increased locker space, superseded the FK4. London Fire Brigade ordered three more limousine BA pumps and nine pump escapes from Dennis Bros, and, from Leyland, its first six-cylinder machines, which were to be limousine 700gpm pumps with BA stowage.

In 1936 the Riverdale Committee, set up to review the fire service in England and Wales, reported that

Right:
East Grinstead's Dennis 'Ace' was a New World pump provided with a tilt to protect the crew, built in 1934 and delivered in February the following year at a cost of £989. *Surrey History Service/Dennis Bros*

Right:
When it was shown at the 1936 Professional Fire Brigades Association exhibition, the prototype Dennis Big 4 was described as having 'a most rakish body', although production models were virtually identical in appearance to the D3 side-valve-engined appliance that it replaced. Driven by a four-cylinder ohv engine through a four-speed gearbox, this model remained in production until 1941. *Surrey History Service/Dennis Bros*

Right:
Aldershot's Dennis Light 6 pump with Braidwood body was new in 1936 and served the town until 1952. It was absorbed into the National Fire Service in 1941, then passed to Hampshire Fire Service on denationalisation in 1948. *Surrey History Service/Dennis Bros*

some 1,500 motor and trailer pumps were available between 770 brigades.

Birmingham had been using Leyland 'gearless' tenders for several years when, in 1936, Leyland built a limousine pump with an automatic gearbox for Edinburgh. There was no clutch and a single lever in the forward position provided speeds up to 25mph, after which the driver pulled back the lever to utilise direct drive from the engine for speeds up to 60mph.

Wembley acquired a Dennis Light 6 limousine pump escape in 1937. A midships-mounted 500-600gpm pump was unusually housed behind the cab rear door, which was vertically divided so that the front half could be opened independently of the back half to give access to the pump. Extending over and behind the rear wheels was a deck area, with lockers beneath, for the carrying of wet hose. The escape mounting brackets were at the rear of this deck and the ladders were recessed into the appliance roof.

In 1937, when Leyland Motors introduced its diesel engine for fire appliances, 7,000 or more such units were already powering commercial vehicles. The six-cylinder 8.6-litre engine developed 100bhp and offered 35% lower fuel consumption than a petrol engine of similar capacity. London Fire Brigade commissioned the first diesel in Britain, a Leyland FT3A limousine breathing apparatus pump, which used 5 gallons of diesel oil an hour under full load. It accelerated from 0 to 40mph in 31 seconds, and was able to deliver 830gpm at 100psi. The next diesel was built by Merryweather in the following year for service at the Royal Naval Armament Depot in Singapore, and Acton Fire Brigade, having been suitably impressed by the London machine, ordered a Leyland diesel.

While limousine pumps were by then familiar in London, in 1937 Merryweather supplied the London Fire Brigade with its first limousine pump escape. The escape was a Merryweather all-steel 50ft model. Birmingham's first limousine appliance was a Leyland pump escape ordered in 1938 and put on the run in 1939.

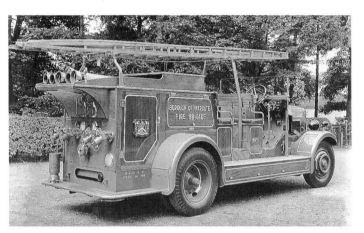

Left:
Margate's Leyland FT3 pump of 1936 was provided with transverse seating and carried a Morris 'Ajax' ladder. The strikingly designed bodywork was of polished natural hardwood and the brigade's FK6 pump escape supplied at the same time was similarly finished. *The British Commercial Vehicle Museum Archive*

Left:
The earlier limousine appliances were van-shaped or square in appearance, but as the decade progressed designs became more flowing, often with a swept back. This Leyland FT pump is a good example of a later limousine bodywork offered with transverse seating for a crew of 11 but no doors. This appliance was delivered to Bolton in 1936 with a Merryweather ladder. *The British Commercial Vehicle Museum Archive*

Right:

Sheffield already had a limousine Dennis 'Ace' rescue tender, but the first 'Ace' limousine pump went to Gellygaer UDC in 1936. The 350-450gpm pump was rear-mounted between the longitudinal seats of the saloon body, which were angled at the back end so as not to obstruct access, in the style of New World appliances with rear-mounted pumps. The hose-reel was mounted inside the forward end of the saloon and the locker accommodation was accessible from inside or outside the vehicle. *Surrey History Service/Dennis Bros*

Left:

John Morris & Sons had been building fire engines on Dennis chassis for 10 years when Glossop Fire Brigade commissioned a six-cylinder 90-100bhp pump from them in 1936. The 500-600gpm pump was rear-mounted and a first-aid set with an 'Ajax' turntable hose-reel was fitted. *By courtesy of 'Fire' magazine*

Right:

In 1937 Ilford Fire Brigade commissioned a forward-control Leyland FK6 limousine pump with a cab for the officer and driver and a compartment behind, with a sliding door, for the breathing apparatus crew. This machine was made distinctive by the fitting of a bell on a forward-projecting bracket over the windscreen on the offside and a searchlight on the nearside. A 'FIRE' sign between these accessories flashed a warning of approach. The main area of the body was open at the rear and housed a 500gpm pump, 50-gallon first-aid tank and hose-reel, as well as foam-making gear, extinguishers, suction and delivery hose, a jumping sheet and a 12ft 'Ajax' ladder. A 35ft ladder and a hook ladder were carried on the roof. *The British Commercial Vehicle Museum Archive*

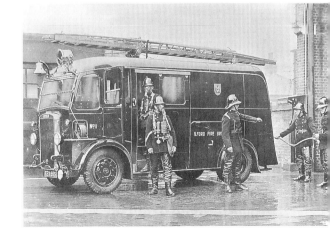

Left:
Kilmarnock took delivery of this Dennis Big 6 pump escape with sweeping limousine bodywork in 1937. *Surrey History Service / Dennis Bros*

In 1939 Coventry Climax announced a new appliance based on an extended-wheelbase Bedford chassis fitted with a 500gpm pump and 100-gallon tank and first-aid set. A four-door cab would accommodate a crew of six, and another four men could be housed behind. By 1941 the company had produced designs for a limousine pump with a four-door cab, on a Fordson 7V chassis, for emergency fire defence. The Government principle of employing a pump with its own engine independent of the vehicle's propulsion unit was maintained. A first-aid set by Charles Winn & Co was installed.

Right:
This Leyland FK6 pump for Rotherham in 1938 displays a rounded back to provide one of Leyland's more unusual designs. *The British Commercial Vehicle Museum Archive*

Left:
The Dennis Light 4 was introduced in 1936 and a number of examples appeared with a cab for the driver and officer, while the rest of the crew were expected to sit in the open on the Braidwood-style body behind. This pump escape was new in 1939 to the joint fire brigade serving the districts of Oakham and Uppingham in Rutland. *Surrey History Service / Dennis Bros*

In 1940 the Dennis line-up was the Light 4 (favoured by rural communities), the Light 6 (500-600gpm), the Big 4 (a 95bhp appliance popular in London), and the Big 6 (650-900gpm, intended for large cities but also popular among brigades with wide areas because of its superior road performance).

What was probably the only fire engine to be bodied by prestige car manufacturer AC (Acedes) Cars of Thames Ditton, was a 1940 Bedford O limousine escape tender that eventually found its way into Scotland's North Eastern Area Fire Brigade in 1948.

As well as Merryweather, John Kerr & Co (Manchester) was providing appliances on Albion chassis. The coachwork of many of Kerr's fire engines was in fact constructed by J. H. Jennings & Son Ltd of Sandbach, including this 1940 limousine pump incorporating a Drysdale pump supplied to Penybont RDC for service at Aberkenfig, Glamorgan.
Norman Tarling collection

When the supply of German imported turntable ladders dried up, this Leyland TLM2A chassis was used as the base for a long limousine pump for

Blackpool Fire Brigade in 1940. The bodywork has often been attributed to Burlingham, but it seems just as likely that the whole appliance was a Leyland product. *The British Commercial Vehicle Museum Archive*

Above:

In 1939 London Fire Brigade ordered 11 Leyland FKT enclosed appliances, with 700-1,000gpm pumps. They were delivered the following year and the rear-mounted pumps were permanently fitted with a four-way collecting breeching as well as a suction inlet for working from open water. The delivery hose was carried in sliding trays. A 2in monitor was mounted over the pump with gearing for elevation and rotation, and a 240ft hose-reel was fitted at the nearside. 'Z-brackets' at the back would accommodate two different types of 50ft escape, or a 40ft extension ladder could be carried as here. The hooks of two hook ladders are clearly visible in this picture. Each appliance carried five breathing apparatus sets and No 10 foam-making branches. The bodywork incorporated four half doors and three rows of seats for a crew of 11.
London Fire Brigade

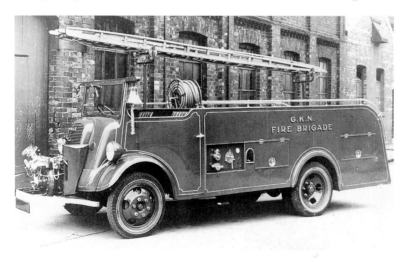

Left:
This Fordson 7V with its front-mounted pump was built in 1940 by John Kerr for the GKN works fire brigade. *Norman Tarling collection*

Wartime Pumps

The Spanish Civil War, which started in 1936, offered lessons on the effects of modern weaponry and the incendiary bombing of towns and cities, and highlighted the important role of the fire service. European peace had been in the balance for some time, and in 1937 the British Government initiated a programme to supply brigades with fire pumps in great numbers. These were different from the appliances already in service because, for convenience of mass production, the pump itself was driven by an independent engine rather than from a power take-off. The idea was not popular with fire brigades, which by then had settled on purpose-built fire engines with pumps driven by power take-off, and provoked comments such as, 'One might just as well fix a diesel engine on a Roman chariot.' The utility design offered an enclosed cab for the driver and officer, while the crew either sheltered on a rear-facing seat or stood alongside the pump. An extension ladder was carried. While they were less sophisticated than their peacetime counterparts, in the event the pumps supplied for wartime use proved themselves well able to pump continuously for days on end.

The heavy unit consisted of a 700gpm pump mounted on a medium lorry chassis and was the chief self-propelled pump of wartime manufacture. A number of coachbuilders undertook the assembly by mounting Leyland-engined Gwynne two-stage pumps on either Austin or Morris Commercial chassis, Ford-engined Sulzer two-stage pumps on either Ford or Bedford chassis, and Ford-engined Tangye single-stage pumps on Ford or Bedford chassis. All had 5½in suction and four deliveries. In the early stages of the emergency period, about 1,000 pumps were built on Bedford chassis and others on Morris Commercial chassis before

these makes were required to fulfil orders for military vehicles. Occasional use was made of other chassis, but thereafter fire engines were constructed on the Austin K4 and Fordson 7V chassis.

The extra heavy unit comprised a Leyland-powered Gwynne single-stage 1,100gpm pump, with a 7in suction and six deliveries, mounted on a Bedford or Austin chassis. It was intended for pumping large volumes in water relays and for supplying equipment needing great quantities of water, such as radial branches and deluge sets. A hundred were to be ordered in 1938.

Creations of the wartime fire service included the dam lorry and the mobile dam unit. The former was generally a flat platform lorry with a dam, or water reservoir, of 500 or more gallons mounted on it. The dam might have been of the steel-framed collapsible or the self-supporting canvas type, or a galvanised iron tank. The purpose of the dam lorry was to convey an emergency water supply to the fire where relatively small amounts were needed or where a water relay was impracticable. If necessary a shuttle service of dam lorries was set up.

If a dam lorry were to tow a trailer pump or have a pump mounted on the platform, it became known as a mobile dam unit. These were used for the first attendance at fires where water supply was likely to be restricted or unavailable and their early deployment could prevent a fire getting out of control. They were also used to patrol areas where the fall of incendiary bombs had been reported, and, because mains water supply was frequently disrupted during air raids, for damping down where the mains had failed. With supplementary equipment these appliances were used by the NFS as aircraft crash tenders. The mobile dam unit was the forerunner of the modern water tender.

Right:
A typical heavy unit on an Austin K4 chassis. *Fire Service Preservation Group*

Trailer Pumps

Trailer pumps were generally mounted on a two-wheel chassis to be drawn behind any suitable motor vehicle and were significantly cheaper than a self-propelled pump. They were provided with their own engine, making them independent of the towing vehicle, and could be manoeuvred into positions where it would be impossible to take a motor pump. These characteristics made them popular in rural areas and with works fire brigades, and their usefulness was perceived to be so great that some anonymous commentator is quoted as saying, 'As the steamer is to the manual, so is the trailer pump to the steamer.'

Simonis claimed to have built the first trailer pump used in England in 1912 for the Enfield small arms factory. Merryweather built its first in 1921, a 250gpm 'Hatfield' for Cairo Fire Department, and advertised its suitability for urban districts. The following year

Tilling-Stevens introduced its Bantam 100gpm trailer pump, recommended for small boroughs, villages and private brigades. It was supplied with pneumatic tyres and, weighing less than 10cwt, could be hand-drawn or towed behind a motor vehicle. At the same time Gwynnes Engineering Co Ltd of Hammersmith was advertising its hand-portable fire engine, a hand-drawn trailer pump able to throw two jets to a height of 80ft.

Leyland was advertising its trailer-mounted 8hp portable pump in 1923, and Baico Patents Ltd of London announced a 20hp trailer pump with a capacity of 200gpm. In 1924 Gwynnes Engineering Co Ltd produced the 'Invincible' trailer pump, a new 100-150gpm machine able to throw two jets to a height of 80ft or a single jet to 130ft.

The 1930s saw a whole range of trailer pumps come on to the market, and these appliances are probably seen by most observers as a symbol of fire-fighting in World War 2. They featured strongly in the

Right:
Curtis's & Harvey of Faversham needed a pump for its explosives loading factory to run on a metre-gauge railway track, so in 1928 a Dennis 200-250gpm trailer pump, popular as a works fire engine, was mounted on a special frame fitted with a horizontal handle for pushing the whole assembly along. The pump itself was powered by a four-cylinder engine provided with an electric self-starter. Two lengths of suction hose and a strainer were carried. *Surrey History Service/Dennis Bros*

Government's emergency preparations and local authorities were encouraged to maintain lists of vehicles suitable for carrying or towing pumps. The Coventry Climax Company had been making motor car engines since 1903, and in 1937 it was recognised that two engines already produced would be suitable for the trailer pumps that the Government wanted in large numbers. So began the 'Godiva' fire pump. The 120/220gpm FSM trailer pump, weighing 448lb, used an 8hp engine originally produced for Swift Motors and was the first pump supplied. The larger FP pump incorporated the 20hp 'P' type engine and 'F' type fire pump capable of delivering 550gpm at 80psi.

The smallest emergency appliance with an independent power unit did not need a towing vehicle. This was the hand-propelled wheelbarrow pump, comprising a single-stage centrifugal pump driven by an air-cooled four-stroke engine, designed for use in factories and public buildings. It had a 2in suction inlet,

and most were provided with one 2½in delivery outlet, although some had two. Pegson and Sigmund both produced 45gpm models and Scammell produced a sled-mounted version in addition to its 65gpm wheelbarrow design.

Light trailer pumps were intended for use in rural and non-industrial areas and comprised a single- or two-stage centrifugal pump, powered by a petrol engine of up to 8hp, on a two-wheel trailer. The pumping unit, with a capacity of between 140 and 175gpm, was generally demountable and provided with two wheels so that it could be moved around away from its trailer. The medium trailer pump comprised a single-stage centrifugal pump, but production was discontinued before the end of the war in favour of light and large models. The large trailer pump fulfilled the need for huge numbers of fire pumps to be built rapidly and supplied in the run-up to World War 2. Capacities went from 430gpm to 500gpm.

Right:
The Coventry Climax FSM pump was dismountable from its trailer and could be manoeuvred by one man with the aid of its rubber-tyred wheels, but carrying handles were provided at each end. *Godiva*

After the war the popularity of the trailer pump diminished, although many remained in service for a good many years. Built-in pumps were a better option and the development of portable pumps for use away from the appliance sounded the death knell for the local authority trailer pump. A JCDD specification for a 400gpm trailer pump was nevertheless produced.

This Dennis large trailer pump produced in 1938 is typical of the units that went on to play a major role in the country's fire defences during the war. *Surrey History Service/Dennis Bros*

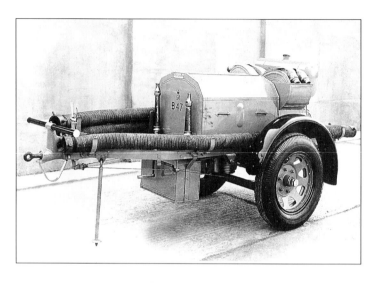

Postwar Pumps

After the war there was much talk of fire engine design, and the need to revert to the best prewar municipal designs was universally recognised by professional fire-fighters. Country-wide standardisation was not seen as a practical option, but fast, limousine appliances were called for. It is noticeable that the early postwar customers for new appliances were industrial brigades, the NFS generally making do with its existing fleet even though its Home Office utility machines were often described as improvised, and even dangerous, lash-ups. Perhaps this is hardly surprising with the promised return of the fire service to local authority control. Many utility appliances of the NFS were, however, rebodied for postwar service in Home Office workshops and by coachbuilding concerns that entered the fire engine market place, as well as by the fire engineering companies.

Leyland Motors concentrated its postwar efforts on commercial vehicle production to the exclusion of fire engines, although its chassis were later used from time to time by other builders. Dennis Bros promptly re-entered the market in 1945 by introducing a new Light 4 appliance, with a wider front axle for increased stability and a 400-500gpm pump. This was followed by the F series in 1946, which lasted until 1979. Merryweather also remained prominent in the field and in 1948 launched the design of a new 1,000gpm pump based on a shortened AEC Regal Mk III chassis with either a 165hp or a 125hp six-cylinder petrol engine.

For urban areas with adequate mains water supply and hydrants, pumps and pump escapes of prewar style were satisfactory, but in rural areas without these facilities they were less than ideal. During the war the assortment of improvised and standard mobile dam units, or water tenders, had demonstrated the usefulness of an on-board water supply of 400 or 500 gallons. From 1948 many local authority brigades therefore sought a modern water tender that would provide the safety and convenience of a limousine appliance with the value of a much bigger water tank than the 60-gallon usually found on prewar machines.

Fire appliances for the denationalised fire brigades were built to a design and specification prepared by the Home Office Joint Committee on Design and Development and, to start with, were centrally ordered for local authorities. The specifications were subject to modification and development but formed the basis of the nation's fire engines until effectively displaced by European standards in recent times.

Two types of water tender evolved from the wartime mobile dam unit: type A, which had no built-in pump but towed a trailer pump, and type B, which did. The essential feature of a water tender was a tank of between 400 and 500 gallons capacity, although developments over the last 25 years have seen midi or compact water tenders with smaller tanks. A portable pump and a 30ft or 35ft extension ladder completed the fundamental requirements. The first water tenders were frequently rebuilt wartime machines and more often than not were of type A. As the availability of chassis and other

Right:

The last years of the war saw many Dodge-chassised fire engines in service with the NFS, many as mobile dam units. After the war it was common for such vehicles to be rebuilt as water tenders, and Hampshire Car Bodies are thought to have produced about 50. This Dodge was converted into a type A water tender for the Isle of Wight Fire Brigade in about 1949, with a portable pump plumbed into the on-board tank.
HCB-Angus Archives

parts improved, the more convenient type B became the generally preferred option, and found favour among brigades with urban, as well as rural, risks.

The JCDD specifications provided for a dual-purpose appliance that could carry either a wheeled escape or an extension ladder. The pump escape was seen as the primary life-saving vehicle, and the biggest advantage of ladder interchangeability was the capacity for a rapid change of role to keep the escape on the run even if its normal carrier were unavailable. The dual-purpose appliance was required to have a water tank of at least 100 gallons and a pump capable of delivering a minimum of 500gpm at 100psi. Lighter and faster appliances, often used in smaller towns or in support of a pump escape, were required to have a pump giving 400-500gpm at 100psi and a first-aid system with at least an 80-gallon tank and one hose-reel. An extension ladder was also an essential requirement. Such appliances were also known as pumps.

The accepted life of a prewar fire engine was 15 years, but the utility appliances built during the war could not be expected to last beyond 10 or 12 years. In his 1949 report, HM Chief Inspector of Fire Services in England and Wales commented that 259 of the 1,667 pumping appliances then in service were between 15 and 20 years old, 49 were over 20 years old, and 776 were wartime machines. Only 12 postwar pump escapes had been built for England and Wales by the end of 1949, with another 70 in prospect by the end of the following year; Scotland had received only one. The 783 water tenders available in England and Wales were, with a few postwar exceptions, makeshift units built during the war on whatever chassis happened to be available, and nearly all of them were overloaded. During 1949 fire authorities were able to contract out the construction of water tenders on bought-in chassis

that were becoming available. No previous fire engineering experience was necessary for firms undertaking this work, the Chief Inspector said, and a central contract for 75 machines was to be placed for completion in 1950.

Whether as a result of that statement or not, engineering firms sprang up to build fire engines on standard or modified commercial chassis, some going on to be market leaders in the field and others falling by the wayside. Among the lasting and successful companies were Hampshire Car Bodies Ltd of Totton and Carmichael & Sons (Worcester) Ltd. The Hampshire company started by rebuilding NFS appliances but soon graduated to the construction of new machines, in the early years mostly on the Commer QX chassis introduced at the 1948 Commercial Motor Show. Motor vehicle chassis for the home market were in short supply and there was a knock-on effect on the construction and delivery times of new fire engines.

In 1950 the City & County of Worcester Fire Brigade was allowed to break from the Home Office specification for new water tenders and commissioned two appliances each with the main pump driven by a power take-off, instead of a light demountable and independently powered pump. In permitting this departure, the Home Office stipulated that when operating in rural areas these appliances had to tow a trailer pump, even though the brigade stated that a demountable pump had seldom been unshipped and used. Of two machines known to have been supplied to the brigade in 1950, a Dennis F7 pump escape was unusual in two ways. First it was bodied by Carmichael rather than Dennis Bros, and second, it was reputedly the first fire engine that Carmichael built. The other machine was a Commer QX, which was reputedly the second.

45

Left:
A user of the Commer QX chassis was Lancashire Fire Brigade with this type A water tender built by Cuerden Motors Ltd of Blackburn in 1950. A 220gpm Coventry Climax wheeled demountable pump was carried at the rear and an auxiliary pump by Charles Winn & Co supplied the hose-reels from a 400-gallon tank. The appliance was equipped to tow a trailer pump. *Roy Goodey collection*

Right:
This Commer QX water tender was a 1951 product of Carmichael & Sons for Angus Area Fire Brigade. Typically 1950s in its design, this model features a cab with four hinged doors. *Carmichael & Sons (Worcester)*

Below:
This pump escape for the Isle of Wight Fire Brigade was ordered in 1950 and delivered in 1952. It represents an example of a Commer QX/Hampshire Car Bodies appliance and features a rear-mounted Dennis No 2 pump and separate hose-reel pump, the controls for which are visible behind the cab entrance. The escape carried was a 45ft Merryweather. *HCB-Angus Archives*

In 1950 two of Merryweather's new 1,000gpm pump escapes built on forward control AEC diesel-powered chassis, based on the Regent III, were delivered to Middlesex Fire Brigade. By the end of the year the AEC/Merryweather diesel had been adopted by four county brigades, six city and county borough brigades, and London had ordered six. The six for Lancashire Fire Brigade incorporated rear-facing crew seats and hose-reels built into the bodywork rather than installed on top.

In 1951 Essex Fire Brigade commissioned the first of four multi-purpose appliances able to run as pump escapes or water tenders. The key element in this role-changing ability was the installation of not one but two water tanks, of respectively 300 and 100 gallons. In water tender mode both tanks would be filled and a demountable pump carried at the rear with an extension ladder aloft. When running as a pump escape only the 100-gallon tank would be filled and the demountable pump removed to allow mounting of a 50ft wheeled escape. A Dennis 350-500gpm main pump was mounted amidships and a 180ft hose-reel was installed on each side. The chassis was a Dodge 125 6-tonner, and although the bodywork was by Arlington Body Builders Ltd of Middlesex, the appliance was fitted out by the brigade itself.

Merryweather was exporting 50ft three-section alloy ladders to Australia and New Zealand in 1949 when Kent Fire Brigade, already looking for something to replace its wheeled escapes, bought two for trials. Experience led to modifications, and the brigade was the first to put a 50ft alloy ladder into service, in 1951, with a view to entirely replacing its wheeled escapes. The alloy ladder weighed about 250lb and cost £200

Above:
By 1952 the Dennis F8, launched two years earlier, was known as the 'Ulster' because of its widespread use by the Northern Ireland Fire Authority, which adopted a model with a 200-gallon first-aid tank. The power came from a Rolls-Royce six-cylinder petrol engine able to propel the 6-ton vehicle at 60mph when fully laden. This example, delivered in May, pre-dates the appearance on Northern Ireland machines of the 'Ulster' designation on the front scuttle, which did not occur until later in the year. *Surrey History Service/Dennis Bros*

Below:
Leeds City Fire Brigade adopted an unusual design for five Dennis F8 pumps built in 1953. While a double cab was provided, with doors for the driver and officer-in-charge, the crew entered from the rear and were accommodated either on a rear-facing seat in the cab or in the open New World body behind. A rear-mounted 500gpm pump, 150-gallon water tank and single hose-reel were fitted to these appliances. *Surrey History Service/Dennis Bros*

Right:
W. H. Goddard, motor body builder of Oadby, is known to have constructed two water tenders on the Commer QX chassis for Leicestershire & Rutland Fire Service, this one in 1951 and another in 1953.
Roy Goodey collection

compared with the £700 for an escape weighing seven times as much. And an appliance with an alloy ladder would fit into fire stations not big enough to accommodate a pump escape. Two years later, Kent became the first brigade to decide that all its wheeled escapes should be replaced by the further modified ladder, now reduced to 45ft. Other brigades followed suit, but elsewhere wheeled escapes survived and it was 1994 before the last escape in service, at Hemel Hempstead in Hertfordshire, came off the run.

The designation of an appliance is influenced by the type of ladder it carries. We have seen that a pump with a wheeled escape was called a pump escape, and by the same logic a water tender with an escape would be a water tender escape. A pumping appliance with a 30ft or 35ft ladder was called a pump, but if it had one of the new 45ft or 50ft ladders it became a pump ladder. Water tenders and water tender ladders were distinguished in the same way. The same generally holds true today, but with hydraulic rescue equipment added to the equipment inventory of some appliances, the word 'rescue' might also feature. The once encouraged

standard nomenclature has now been replaced by what sometimes appears to be a competition among brigades to outdo one another by coming up with longer and more impressive designations.

Dennis Bros introduced the F12 pump and pump escape in 1950 to replace the F7 model. The new appliance had a wheelbase of 12ft 6in, 1ft shorter than its predecessor. By the end of the year 75 English brigades had ordered at least one F12, Middlesex being the best customer with 21. Before the model was discontinued in 1959 Dennis had produced 336 F12s.

Karrier Motors introduced a new model known as the Gamecock at the 1952 Commercial Motor Show; two versions were available with either a 9ft 7in or an 11ft 9in wheelbase, both powered by an underfloor six-cylinder engine. This proved to be a popular chassis, particularly for Carmichael-built smaller appliances.

The postwar AFS initially relied on wartime appliances but a large fleet of purpose-built fire appliances was established in the mid-1950s with the intention that some would be deployed in mobile fire columns of 144 vehicles to reinforce hard-pressed

Right:
Kent Fire Brigade commissioned a new Leyland Comet pump ladder built by Windovers Ltd of Hendon in 1951. The petrol-engined appliance had a rear-mounted Dennis 350-500gpm pump and a Charles Winn 50gpm hose-reel pump fed from the 100-gallon water tank. The rear gallows for the Merryweather 50ft ladder was modified to slide up and down enabling the ladder to be carried horizontally on the appliance, reducing the overall height so that it would fit in the fire station, but dropped to a level enabling removal when required.
Kent Fire Brigade Museum

Left:
This 1954 Dennis F12 was delivered to run as a pump ladder from new, carrying a 45ft Merryweather alloy ladder instead of the more usual wheeled escape of the time. While following the general lines of the F7, this model can be distinguished by the less-than-full-width locker behind the pump bay. Note the rolling mechanical siren on the roof at the nearside. *Surrey History Service/ Dennis Bros*

Right:
Another distinctive Dennis F12 was this 1954 pump for Salford, which had no provision for carrying a ladder at all when it was new. Gantries were added later, enabling it to carry a 45ft alloy ladder and run as a pump ladder. *Surrey History Service/ Dennis Bros*

Left:
The body style adopted for this Bedford S pump escape of Scotland's Central Area Fire Brigade in 1952, with its rounded roof-line and inset bells, is unmistakably by Alfred Miles Ltd of Cheltenham. Popularly known as Big Bedfords, the forward-control S range appeared at the 1950 Commercial Motor Show and provided a favoured alternative to the Commer. Alfred Miles used aluminium alloy for its fire engine construction with a body and cab weighing only 11½cwt. *Vauxhall Motors*

local services anywhere in the country. These vehicles happened to be painted green, but not because they were under military control, and the emergency pump, or self-propelled pump as it was described, became affectionately known as the 'Green Goddess'. It is these emergency pumps, taken back into Home Office depots on disbanding of the AFS in 1968 and now 45 years old, that are still brought out in times of crisis and sometimes crewed by military personnel, encouraging the misconception that they are Army fire engines.

There were two versions of the emergency pump, the earlier type built on the Bedford SHZ rear-wheel-drive chassis with a 400-gallon water tank and 180ft hose-reel provided on each side at the back. The later type used the Bedford four-wheel-drive RLHZ fitted with a 300-gallon tank and hose-reels mounted centrally. Both incorporated a rear-mounted Sigmund 900gpm pump and carried a Coventry Climax FWP 300gpm light portable pump in a special locker with a sliding ramp; 1,600ft of rubber-lined delivery hose and

a 35ft alloy ladder were included in the inventory of equipment.

Petrol-engined fire engines were still by far the most popular at this time, but diesel was making a gentle impression. Tilling-Stevens had become part of the Commer organisation in 1951 and developed the TS3 diesel, which appeared at the 1954 Commercial Motor Show. Meanwhile a diesel-powered Dodge type B water tender built by Herbert Lomas Ltd was displayed at the 1954 IFE/CFOA conference. This appliance incorporated a rear-mounted Dennis pump and a midships-mounted Hathaway high-pressure pump. Also in 1954 Nottinghamshire Fire Brigade commissioned a new water tender escape built by Wilsdon & Co Ltd of Solihull on a forward-control Leyland PD2/10 chassis with 13ft 6in wheelbase. The engine was a six-cylinder diesel developing 125bhp at 1,800rpm. A Dennis 500gpm pump was installed at the rear and a Coventry Climax featherweight pump was carried in a locker. The 50ft wheeled escape could be replaced by a 35ft alloy extension ladder as necessary.

Left:
The F8 had been an all-Dennis machine until 1954 when Alfred Miles exhibited an appliance it had built on the Dennis chassis with a Rolls-Royce 120bhp engine. This was a true water tender with a 400-gallon tank and portable pump, yet it remained only 6ft 6in wide and 9ft high. *Surrey History Service/Dennis Bros*

Right:
Another Miles-bodied appliance was this 1954 Bedford A water tender of the Kesteven (Lincolnshire) Fire Brigade. The Bedford A series was introduced in 1953.
Roy Goodey collection

Below:
Merryweather launched its first Marquis in 1954, built on the AEC-Maudslay chassis and designed as a light appliance for both urban and rural fire-fighting. In 1955 Surrey and Dewsbury fire brigades each commissioned the water tender version, while the South Eastern Fire Brigade in Scotland took one as a water tender and this one as a pump.
Author's collection

The prototype Dennis F24 was seen at the end of 1955. The chassis had a wheelbase of 10ft 9in, longer than the F8 but shorter than the F12. Motive power was from a Rolls-Royce B80 eight-cylinder petrol engine, and the Rolls-Royce-built automatic transmission had a two-speed transfer box and power take-off. To those familiar with the production models, the striking aspect of the prototype was its F12-style cab.

This history so far has demonstrated that in the 10 years since the war, only Dennis had undertaken chassis-building, fire engineering and bodybuilding to produce complete fire engines. Fire engineering companies and coachbuilders had used commercial chassis — predominantly the Commer QX and the Bedford S, with AEC favoured by Merryweather.

Dennis and Merryweather were the long-established names in fire engineering that had survived the wartime disruption. Of the new entrants into the arena,

Hampshire Car Bodies, Carmichael and Alfred Miles were the undoubted market leaders, the latter making use of aluminium for its appliances when the norm was oak and ash framing with steel, and later alloy, panelling. A representative selection of the rest has been illustrated, but there were others, of which perhaps Wilsdon & Co produced the greatest number of machines. And so it was the activities of these firms that enabled the chairman of Dennis Bros to comment in his 1955 annual report that the postwar re-equipment demand for fire appliances was almost satisfied.

The Merryweather Marquis II on an AEC diesel-powered chassis was announced at the end of 1956, and an early customer for a new water tender was Merthyr Tydfil Fire Brigade. The Marquis III, with wrap-round windscreen, was announced in the following year and provided a synchro-mesh gearbox as standard with a constant mesh power take-off.

Right:
Lancashire County Fire Brigade took delivery of this Carmichael-bodied Karrier Gamecock water tender escape in 1956. Note the pump controls and deliveries in the rear locker over the wheel arch to facilitate convenient operation without slipping the escape.
Carmichael & Sons (Worcester)

Below right:
In 1956 the automatic Dennis F24 became available together with the alternative F25 five-speed manual gearbox appliance powered by a six-cylinder 114hp engine. Both were offered with a choice of 500-600gpm or 900-1,000gpm main pumps, in dual-purpose, water tender or multi-purpose layout. Water tanks with capacities of 100, 250 or 400 gallons, or a two-compartment tank for water/foam, were available, and different styles of cab and locker doors were offered. Norwich City Fire Brigade was the first to put an F24 on the run, and this F25 water tender was delivered to Somerset Fire Brigade in 1957. *Surrey History Service/ Dennis Bros*

Pilkington Bros works fire brigade took delivery of what must have been one of the last, if not the last, Braidwood-bodied appliance built for service in the UK. This 1956 Dennis F2 pump ladder is unusual for that reason alone, but the 45ft ladder provided as part of the original specification for this open appliance surely makes it unique. *Surrey History Service/Dennis Bros*

The Royal Air Force domestic appliance, for tackling fires not involving aircraft on ground bases such as ammunition and maintenance establishments, was under development by Alfred Miles in 1957 using the Bedford S chassis. A Coventry Climax 500gpm pump and two water tanks with a combined capacity of 600 gallons were fitted. These were the first purpose-designed domestic tenders built for the RAF, which had previously relied on a Karrier Bantam flatbed appliance with chemical foam system, the wartime Austin towing vehicle and trailer pump, and a modified Bedford OYC water carrier.

In 1958 Leyland Motors announced that it was once again to manufacture fire engines. In September the Leyland Firemaster chassis was introduced with a 150hp diesel engine to power the appliance; it had a 12ft 6in wheelbase and low centre of gravity, and was fitted with a front-mounted 900-1,000gpm pump and a first-aid pump delivering 75gpm at 500psi. The first chassis was produced for Manchester Fire Brigade to be bodied by Carmichael.

The completed vehicle was not seen until 1959 and Manchester went on to acquire only two more Firemaster pump escapes, bodied by Cocker of Southport, and an emergency tender. Essex and Glasgow had the appliances shown in the accompanying illustrations, and the only other Firemasters were built as turntable ladders.

The Dennis F28 appeared 1959 and, at 7ft wide, was a smaller machine than the F26, with a choice of either a 12ft 3in or a 10ft 9in wheelbase. It was driven by a Rolls-Royce six-cylinder engine developing 150bhp, and the Dennis No 2 pump was installed. In 1960 Dennis Bros announced a new pumping arrangement for its appliances, with a No 2 pump coupled to a Hathaway pump for medium/high-pressure output. The single-stage Hathaway pump would provide the first-aid pumping facility.

In 1961 Essex Fire Brigade put two Leyland Firemaster composite appliances on the run at Colchester and Grays. Built by David Haydon Ltd of Birmingham and known as PESTs (pump emergency salvage tenders), they had a 1,000gpm front-mounted pump and a high-pressure pump for the two hose-reels fitted with fog guns. A 300-gallon water tank was installed over the back axle. Behind the crew cab, a work space with two 6ft benches was incorporated for BA stowage and servicing. An Epco set, cutting equipment and salvage gear were carried in addition to a range of fire-fighting equipment and an extension ladder. *The Wardell collection*

Above:

Glasgow had two Leyland Firemaster pump escapes, one by David Haydon in 1960 and this one by Cocker in 1961. *The British Commercial Vehicle Museum Archive*

Right:

This Bedford S water tender ladder was built by Carmichael for Doncaster Fire Brigade in 1959. It displays the revised radiator grille in use from 1957. *Carmichael & Sons (Worcester)*

Left:
The Dennis F26 was seen in 1959 and came with a choice of engines: the Rolls-Royce 160bhp B80 or the 195bhp B81. On a 12ft 3in wheelbase with a width of 7ft 6in and with a 400-gallon tank, it was promoted as a replacement for the F12 and F15 models. Either the Dennis No 2 pump, giving 600gpm through its two delivery outlets, or the larger four-delivery 950gpm No 3 pump could be installed. With the latter midships-mounted, as on this Huddersfield pump escape, three deliveries each side were provided.
Surrey History Service/Dennis Bros

Right:
A unique appliance in the early 1960s was this Alvis Salamander/ Pyrene water tender of the Army Fire Service, looking very much like the RAF Mk 6 crash tender described later but without the roof monitor. This machine was based at the Central Ammunition Depot, Kineton, and was fitted with a 900gpm pump and 700-gallon water tank. *Crown copyright*

The Bedford TK chassis was introduced in 1960 and promptly became a firm favourite with fire engine builders and their customers alike. Carmichael announced the production of a new water tender on the Bedford TK chassis in 1961, using standard Bedford panels for cab and doors with an alloy body and glass-fibre roof. Alfred Miles exhibited a water tender on the Bedford TK chassis at the 1961 IFE/CFOA Edinburgh conference.

Ambulance-builder Busmar Ltd, of Blackpool, made its first appearance at an IFE/CFOA conference in 1963 when the company displayed a dual-purpose appliance built on a Bedford TKEL chassis, with an ash frame, diamond-patterned stucco aluminium panelling and Dover roller shutters. In 1964 Warrington Fire Brigade commissioned a multi-purpose appliance built by Busmar on a Bedford TK 6-ton chassis to fulfil the roles of pump, water tender, foam tender, emergency tender and salvage tender.

The year 1963 saw the introduction of the Albion-chassised Firechief by Carmichael & Sons (Worcester). The chassis was a special conversion of the 7-ton Albion Chieftain Super Six; a Leyland 6.54 diesel provided the motive power and a 900gpm pump and 400-gallon water tank were fitted. The first built went for export.

London Fire Brigade ordered 31 diesel-powered Dennis F106 appliances in 1965, at a cost of over £190,000. Four of them were to be dual-purpose machines with Dennis 900gpm centrally mounted pumps and 100-gallon tanks, while the rest would have rear-mounted 900gpm pumps and 300-gallon tanks. Apart from four designed to carry 45ft ladders, even these would be fitted for wheeled escapes. All were to be fitted with two-tone horns as part of the effort to standardise warnings in the Greater London area.

The normal-control Bedford TJ range was introduced at the 1958 Commercial Motor Show, and the 5-ton J4 and 6-ton J5 models became popular as fire engine chassis after conversion to forward-control. This conversion of a J5 water tender was engineered by HCB with the approval of Vauxhall Motors, except for a driving position that resulted in the gear lever being some 24in behind the driver. The appliance, with HCB bodywork and a Hayward Tyler pump, went on the run with Lancashire County Fire Brigade in 1959. *HCB-Angus Archives*

This 1967 Bedford TK water tender by HCB-Angus for East Sussex Fire Brigade is typical of many built by the company in similar style until the TK was discontinued in 1984.
Roger C. Mardon

The Austin FFG chassis did not provide the base for many fire engines, but Scotland's Western Area Fire Brigade took delivery of this pump completed by HCB Engineering in 1963. It was equipped with a 350gpm Godiva pump and carried 100 gallons of water and a 30ft ladder. *HCB-Angus Archives*

Left:
In 1962 Carmichael & Sons built this water tender ladder on a Bedford J chassis, converted to forward control, for Worcester City & County Fire Brigade. The appliance was powered by the Bedford 300cu in engine and was fitted with a rear-mounted Gwynne combined high/low-pressure pump. A Merryweather 45ft alloy ladder was carried on the roof. *Carmichael & Sons (Worcester)*

Right:
This Bedford J4 pump escape was a 1962 HCB Engineering product of all-metal construction, which drew heavily on the company's commercial bodybuilding experience of the 1950s. This Durham County machine was fitted with a 350-gallon water tank and carried a 50ft escape. Note the side controls of the rear-mounted Godiva pump giving the operator easy access without the need to slip the escape.
HCB-Angus Archives

Left:
Burnley Fire Brigade put this Leyland Comet Super Six water tender on the run in 1963. It was engineered by John Morris & Sons and fitted with a 500-600gpm main pump and carried a 250gpm portable pump.
Author's collection

Above:

In 1964 Kent Fire Brigade invited its members to submit ideas for incorporation in a new appliance, and some 1,250 responses were received. Realistic suggestions were taken up and in 1965 the result was the K2 water tender. Built by HCB-Angus on a Commer VAKS petrol-engined chassis, the appliance had a 500gpm rear-mounted pump and a 400-gallon water tank. The bodywork was unpainted alloy, except for a red front, and accommodation was provided for a crew of six. The machine could be run as a water tender, water tender ladder, pump, pump ladder or pump salvage tender. Interesting features were the nearside slide-out BA rack at a natural height for dressing and a rear ladder built-in for roof access. Suction and long gear were carried in compartments, with access from the rear, immediately above the nearside rear wheel arch. Hose-reels, with manual rewind mechanism, were in low compartments behind each rear wheel. After a tour of the county, the first K2 went on the run at Dartford. *Roger C. Mardon*

Right:
ERF exhibited its first fire engine chassis at the Commercial Motor Show in 1966. The model 84PF was powered by a Perkins V-8 diesel and designed for pumping appliances, the first of which was bodied by HCB-Angus and destined for Newcastle and Gateshead Fire Brigade. A Coventry Climax 750gpm pump was rear-mounted, but with side controls to give the operator ready access without having to slip the escape, and a 100-gallon first-aid tank and two hose-reels were fitted. Crew seats faced to the rear. Later versions, designated 84RF, would become available with a Rolls-Royce petrol engine. *Author's collection*

In 1966 Dennis Bros exhibited the new Rolls-Royce-engined F36 dual-purpose appliance built for Coventry Fire Brigade. This appliance was visually striking because of its unique yellow colour, developed for speedy recognition on the crowded roads of the day. It was designed as a combined fire-fighting and rescue vehicle with a pair of Dean 180ft hose-reels that not only delivered water for fire-fighting but also supplied air for power tools from a built-in compressor. It was possible to operate air-tools from one hose-reel while providing a first-aid jet from the other. A Dennis No 3 900gpm pump was rear-mounted, with side controls,

and a 50ft steel escape was carried. The hose-reels were in independent lockers, but otherwise one roller shutter each side gave access to a subdivided through locker housing all equipment apart from breathing apparatus, which was stowed in the crew cab.

In 1968 turbocharging was on the agenda of truck manufacturers and Ford was among the leaders in introducing a turbocharged diesel for its D600 range. HCB-Angus had already used this chassis and was alive to the greater flexibility of performance, less gear changing and improved acceleration likely to be obtained in the future.

Left:

The Merryweather Marquis Series 7 range was launched in 1966 with power-steering and rear-facing crew seats. The midships- or rear-mounted Merryweather MA 1-55 625gpm pump was standard, but a choice of tank capacities and ladder mounting arrangements was available. The first, with a 400-gallon tank, was delivered to Leicester the following year. While the vehicle was finished in red, the 50ft steel escape was yellow to increase its night-time visibility. *Author's collection*

Left:

In 1967 Gloucestershire Fire Service acquired the first of its Ford D600 petrol-engined appliances built by HCB-Angus. This was of innovative design with the crew compartment forming part of the main bodywork, accessed via jack-knife doors on each side, and the standard tilt Custom-cab retained. A Coventry Climax UFP pump was installed at the rear, a 400-gallon tank was fitted, and provision was made for carrying a portable pump. Hose-reels and pump were protected from freezing by hot air circulation. Either a Lacon 45ft ladder or a 35ft extension ladder could be mounted above, enabling the appliance to run as a water tender ladder or as a water tender. *Author's collection*

The Jaguar-powered Dennis D series, the Dennis F108, a manual-transmission appliance powered by a Perkins V-8 diesel, and its automatic equivalent, the F109, all appeared in 1968. The Merryweather Marksman range of appliances also made its debut, and in 1969 Lancashire Fire Brigade ordered 14 Marksman appliances on Ford turbocharged diesel-engined chassis.

In 1968 Glasgow Fire Service used the German Deutz 150D 14FL chassis for four pump escapes bodied in Edinburgh by SMT Sales & Services Ltd. Each had a 1,000gpm rear-mounted pump with side controls and a 350-gallon water tank, and carried a 50ft wheeled escape. This was the first departure from the use of British chassis by any brigade since the war, but at the time it did not seem to herald the extent to which European makers would penetrate the British fire engine market in the years to come.

According to statistics produced jointly by the Institute of Municipal Treasurers & Accountants and the Society of County Treasurers, at the end of 1969 the 131 fire brigades of England and Wales had in service 2,955 pumps and water tenders and another 2,829 portable and trailer pumps. London had 215 pumping appliances, none of which were water tenders, while all of Warwickshire's 37 were water tenders.

The Dodge K850 was a specially developed fire-fighting chassis in the K500 series of trucks plated at 11½ tons gross vehicle weight, which first appeared at the beginning of the 1970s with an overall width of 7ft 6in. An HCB-Angus-bodied example at the 1971 IFE/CFOA conference was powered by the Perkins 185bhp V-8 diesel engine, offered in place of the standard unit.

In 1972 an Institution of Fire Engineers scholarship paper on accidents to firemen had demonstrated that 10% of fire service accidents occurred in or on appliances, and that half of those involved getting into or out of the vehicle. Loughborough Consultants Ltd, established in 1969 by Loughborough University of Technology, had also undertaken a major project on fire engine design, and crew safety moved to the top of the agenda. Of course there was no instant design for the perfect fire engine. A purpose-built fire engine chassis offered the opportunity to build within the parameters of good design, while the use of a commercial vehicle chassis was a compromise offering advantages and disadvantages. Commercial chassis were cheaper and probably, with local spares availability and tilt cabs for ease of access for servicing, offered lower maintenance costs. On the other hand they were not inherently designed for the rapid entry and exit of a full fire crew and their equipment.

Right:
Dudley County Borough Fire Brigade commissioned this Albion/Carmichael Firechief water tender in 1969. The crew cab was entered via a sliding door. *Carmichael & Sons (Worcester)*

Right:
In 1969 Dennis Bros had in build for West Riding Fire Brigade 35 F46 appliances finished in brilliant white, one of which was exhibited at the IFE/CFOA conference. *Author's collection/West Riding Fire Brigade*

The local government reorganisation of 1974 created new fire authorities and in some areas threw together in one fleet the many and varied appliances of the constituent brigades. This highlighted the availability of chassis manufacturers and bodybuilders and the ways in which the two combined to make a fire engine. Diesel-powered appliances were not new but had been given added impetus with the introduction of the turbocharged engine. Rolls-Royce and Jaguar engines were no longer confined to niche manufacturers and HCB-Angus had made these power units available in Bedford fire engines.

The mid-1970s saw a flurry of activity with new manufacturers coming on to the scene, as well as new appliances from established manufacturers to satisfy the demand for greater crew safety. In 1975 Anglo Coachbuilders built two water tender ladders on Ford D series chassis for Surrey Fire Brigade, and the next year the company exhibited at the IFE/CFOA conference for the first time.

The Shelvoke & Drewry chassis made its debut in the fire engine market in 1975 when a water tender jointly designed with Carmichael & Sons, and designated the CSD, was exhibited at the Interfire exhibition.

Eagle Engineering Co Ltd of Warwick entered the UK domestic fire engine market in 1976 and announced orders for five Dodge K850 machines with Godiva UMP Mk 50 500gpm pumps for Lothian & Borders Fire Brigade and four Bedfords for Cornwall. Stonefield machines were also first introduced to the fire service in 1976 when Anglo Coachbuilders exhibited the Ayrshire-built 6x4 chassis at the 1976 IFE/CFOA conference.

G & T Fire Control Ltd of Kent was already known for its turntable ladder rechassis conversion work when in 1976 the company introduced the 'Attack' light pump built with a 150-gallon tank on the Ford A0610 chassis. The G & T 'Advance', with a 300-gallon tank built on the Dodge G08 chassis, was launched in 1978.

Right:
Developed as a result of a report into fire engine design by Loughborough Consultants Ltd, the Chubb Pacesetter, built on a Reynolds Boughton chassis, was launched at Interfire 1975 and constituted Chubb's entry into the general fire engine market. The rear-engined appliance was powered by a General Motors V-6 236bhp diesel engine with Allison automatic transmission. The pump was a front-mounted Godiva multi-pressure unit with ratings up to 1,000gpm, and the 2,000-litre water tank was of glass-reinforced plastic. A six-man cab was provided with a low-level nearside entrance 4ft wide and with a power-operated door. Suction was carried at low level in lockers between the wheels, and low-level ladder access was also provided. Two Pacesetters were sold in the UK, one to Merseyside Fire Brigade and the other to the Fire Service Technical College, as it then was, at Moreton-in-Marsh. *Roger C. Mardon*

Dennis, having been acquired by Hestair four years earlier, introduced the R series of appliances in 1976 to meet calls for an inexpensive specialised vehicle offering good performance without prejudicing quality standards. Apart from petrol or diesel engine, manual or automatic transmission, and single or multi-pressure pump options, the manufacturers offered a wood or steel framework. The first Dennis R pumping appliance went to Clwyd Fire Service.

HCB-Angus introduced its new CSV crew safety vehicle in 1976, which had been produced after research with the Cranfield Institute of Technology and testing at MIRA. The cab, with interior design by Ogle Design Ltd of Letchworth, was promoted as able to withstand crush and impact loads far greater than any known existing designs. The first production model was built on a diesel-powered Bedford KG500 chassis.

The steel-cabbed Dennis RS appliance, another machine with Ogle input to the cab design, was announced in 1977 as the intended replacement for the R series. The first order was from Greater Manchester Fire Service for 14 RS131 water tender ladders powered by Perkins V-8 540 diesel engines with Allison automatic transmission, delivered in 1979. The first Dennis RS in the Republic of Ireland entered service at Sligo the same year.

The Dodge G13 replaced the K850 fire engine chassis in 1978. The new 13-ton appliance was developed partly with Cheshire Fire Engineering and a Hi-Line tilt cab featured in the design. Excluding VAT, it cost £10,400 with a Perkins V-8 5540 engine. Also in 1978 Trailex, the special vehicle division of Contact Displays Ltd, Bournemouth, built two water tenders for Surrey on 1975 Ford D chassis that the brigade already had. The company had built breathing apparatus control units for West Sussex, but this appears to have been its only venture into pumping appliances.

Right:
This example of a Bedford TKH/CB-Angus CSV was one of three built for the Ministry of Defence in 1978 as convoy support vehicles to travel with and protect nuclear weapons convoys. They were equipped with the normal rear-mounted 500gpm Godiva pump but carried 500 gallons of pre-mix foam instead of the standard 400 gallons of water of a local authority water tender. No roof-mounted ladders were carried, but an additional lighting mast was installed. *HCB-Angus Archives*

Left:
Cheshire Fire Engineering Ltd (CFE) of Winsford, a wholly owned subsidiary of ERF Holdings, came on the scene in 1977 with Bedford KG and Dodge K850 water tenders. This Bedford was built for Northamptonshire in 1980.
Roger C. Mardon

Left:
In 1979 HCB-Angus developed its high-strength cab with unusually wide doors for easy access to the front and rear of the crew compartment. Known as the 'Moore's front-end', after Moore Plastics of Slade Green, Kent, who undertook the moulding for HCB-Angus, the HSC model was based on the Bedford TK chassis, but was available at lower cost than the CSV. The appliance pictured is a 1980 demonstrator with a 1,818-litre tank, Godiva rear-mounted pump and 13.5m ladder.
HCB-Angus Archives

The 7ft-wide Dennis DS with tilting steel safety cab was introduced in 1981 for rural areas as a replacement for the D series, and the first went into service with Hereford & Worcester Fire Brigade. Dennis had developed the larger SS chassis, also with tilting steel safety cab, by 1982, and London Fire Brigade ordered 40 SS131 models in that year. The cab could be tilted hydraulically by one man to an angle of 60° to give access to the engine for maintenance.

The decline of the British truck industry in the 1980s had a knock-on effect on fire engine manufacturers by changing the chassis available on which to build. Dennis and Dodge remained as constant players, but ERF, Ford and Shelvoke & Drewry left the fire appliance market in the early 1980s, with Bedford closing in 1986. They were replaced by Leyland and, from mainland Europe, by Mercedes-Benz, Scania and Volvo. To compound the difficulties, this coincided with a tightening of local authority budgets, with brigades looking for cost savings, often translated into less money for new vehicles.

ERF was unable to sell CFE after some two years of trying, so closed it down in 1982. Saxon SVB, a private limited company founded by former CFE staff,

promptly appeared and maintained responsibility for spares and servicing of ERF and CFE appliances, as well as producing new machines. Since 1984 the company has been known as Saxon Sanbec, after the old name for Sandbach. Mountain Range Ltd of Crewe had been converting commercial vehicles into fire appliances for the export market for about three years when it opened a new factory in 1982.

In 1984 Chubb sold its fire engine business to Gloster Saro, which was itself later sold to Simon Engineering. Merryweather, having already left London for Wales, suddenly moved to Plymouth in 1984 before being closed by down by parent company Siebe Gorman.

In 1985 Hestair Dennis reorganised and Dennis Specialist Vehicles was formed to concentrate on production of specialised chassis, buses and coaches as well as fire engines. John Dennis Coachbuilders Ltd was established as an independent company by former Hestair Dennis managers and engineers, originally to carry out bodywork repairs and servicing. However, it was not long before the company's first order for new appliances was received from Staffordshire, followed soon after by one from Cleveland.

Right:

In 1980 North Yorkshire ordered an unusual water tender ladder from HCB-Angus on a Bedford TK 4x4 Pathfinder chassis produced by United Services Garages (Portsmouth) Ltd. The extended chassis incorporated a Bedford 500 156bhp engine. Fire-fighting provision included a Godiva pump, 400-gallon water tank and Angus 45ft ladder. *HCB-Angus Archives*

Left:

HCB-Angus built only a handful of vehicles on the Shelvoke & Drewry chassis. In this example, one of five water tender ladders for Surrey Fire Brigade ordered in 1981, the chassis/cab is a product of Shelvoke & Drewry while the rear bodywork and fire engineering were undertaken by HCB-Angus. A water tank of 1,818 litres and a Godiva pump were installed. *HCB-Angus Archives*

Below:

This Bedford TKG water tender for Leitrim County Fire Brigade in Ireland represents an unusual design for Angloco, who built it in 1981. *Gordon Baker*

Walter Alexander & Co (Belfast) Ltd had been building fire engines for 12 years or more when in 1986 the company produced its first appliances to cross the Irish Sea for Lothian & Borders Fire Brigade. The vehicles were powered by the Perkins V-8 540 diesel engine, and the Godiva UMPX 50 pump was installed. The company also achieved orders from Dumfries & Galloway and Northamptonshire fire brigades in the same year, all vehicles being based on the Dodge G13 chassis.

Locomotors had built crash tenders for the British Airports Authority and some demountable pods for London, but its first local authority pumping appliances were built on Dennis RS chassis for Berkshire in 1986.

Following the success of its Mercedes-Benz/Polyma emergency tender, Norfolk Fire Service bought one water tender from the same manufacturers in 1986, to be followed by another 13 over the next 2½ years. In 1987 Excalibur CBK Ltd of Burslem completed an order for five compact water tenders for Dumfries & Galloway Fire Brigade and secured an order for five more. Built on the Dodge G08 chassis with 3m wheelbase, the crew cab was integrated with the rear bodywork and a Godiva UMPX 50 500gpm multi-pressure pump was rear-mounted. A portable pump was also carried, but the water tank capacity was reduced to 220 gallons. Only one hose-reel was provided, but it was 240ft long and equipped with electric rewind.

Reynolds Boughton (Devon) Ltd introduced its Brigadier water tender, specially designed for county brigades, at the Fire 87 exhibition. This had been produced following two years of research with brigades into what the ideal fire engine should be like, leading to a vehicle with a low centre of gravity and a safety cab with an improved environment for the crew. In a way reminiscent of the Chubb Pacesetter of 12 years earlier, also on a Reynolds Boughton chassis, the six-man steel-framed cab incorporated low-level entry through a nearside air-operated door. The appliance was evaluated by Devon Fire Brigade and ran as Exeter's first-away machine, answering 150 calls and covering over 2,000 miles. It received favourable comment but no more were built and the prototype was sold to Mobil Oil, Coryton. After the Brigadier, Reynolds Boughton came on to the local authority fire engine scene in 1989 with an order from West Sussex Fire Brigade for four Dennis SS239 water tenders, and a three-year contract to supply seven machines to Berkshire.

While the first UK appliance on a Scania chassis was a turntable ladder for Grampian Fire Brigade in 1981, the first Scania water tender for a British brigade was delivered in 1986. Six such machines were ordered by Strathclyde Fire Brigade from Fulton & Wylie on the G82M 4x2 chassis powered by a Scania 7.8-litre 190bhp diesel engine. Godiva UMPX 50 pumps were installed.

The first Volvo pumping appliance to enter service with a UK brigade was built by Fulton & Wylie on the FL6.14 chassis for Clwyd Fire Service in 1987. This vehicle, with a 4m wheelbase, was powered by the TDF61F intercooled engine through an Allison MT 643 automatic gearbox. The crew cab, integrated with the standard Volvo safety cab, was designed by consultant Dawson Sellar of Kirkcudbright.

Suffolk Fire Service replaced its vehicles on a 15-year cycle, and after Bedford went out of production in 1986, in common with many other brigades, had to decide which chassis to adopt in its place. Again in common with many brigades, it chose the Volvo FL6.14 and in 1989/90 took delivery of water tenders from HCB-Angus. By 1990 Volvo was claiming 40% of the British fire engine market.

In 1974 the new Greater Manchester Fire Brigade had established a vehicle replacement policy under which pumping appliances were given a life of eight years and special appliances 15 years to reflect their lower level of use. In 1976 the brigade's pumps were averaging just under 6,000 miles a year and turntable

Left:
In the mid-1980s Timoney, an emerging manufacturer of County Meath in Ireland, was developing a fire engine with independent suspension based on an armoured personnel carrier design. Nineteen were built, but they were expensive. In 1985 Kildare County Fire Brigade commissioned this water tender built by Timoney on its own chassis. *Gordon Baker*

Right:
This Dodge G13/Saxon water tender was commissioned by Avon Fire Brigade in 1986. *Roger C. Mardon*

Right:
This 1990 Bedford TL/Fulton & Wylie water tender ladder of the Defence Fire Services served at Marchwood military port in Hampshire. *Roger C. Mardon*

ladders no more than 1,000. Increased calls, training and off-station duties, including fire prevention inspections and exercises, resulted in increased mileages and by 1990 the brigade's pumps were averaging nearly 8,000 miles a year. Turntable ladder mileage had gone up dramatically to 5,000. The seven emergency salvage tenders had gone from under 5,000 to nearly 11,000 miles each, hydraulic platforms from under 3,000 to 7,000, and foam tenders down from 5,000 (in 1979) to about 4,500. These figures no longer justified the differential replacement ages and the brigade proceeded on the basis that an average life of 10 years could be expected for all operational appliances.

In 1990 the European Committee for Standardisation (CEN) set up Technical Committee 192 to establish European standards for fire-fighting equipment, effectively to supersede the JCDD specifications that had served the British fire service for over four decades. This was not because Europe needed a standard fire engine or a standard hydrant, indeed probably quite the opposite, but because a fundamental principle of the European Union provides for free trade, which is apparently restricted by the existence of national standards.

Traditionally equipment was held in tailor-made fittings on bulkheads or shelves, but this was inconvenient when equipment changed, and some flexibility was obtained through the use of adjustable bollards or shelving. The Manual Handling Regulations, in force from 1993, influenced the design of equipment stowage on fire engines, and slide-out removable trays, slide-and-tilt drawers and hinged swing-out panels all offered greater versatility as well as safer access. Regulations governing working at heights effectively outlawed the appliance roof as a working area, and innovative ladder gantries, such as the AS Fire & Rescue Equipment beam gantry, enabled the shipping and unshipping of ladders from ground level. Continued pressure to save money forced reconsideration of traditional practices, and the standardised design and configuration of pumping appliances within a brigade was one thing to suffer. So-called compact water tenders were introduced for less busy stations, sometimes because they offered an operational advantage but also because they were cheaper.

Over the years fire engines have had to become secure against crime. Security of both crew and equipment became a serious matter, with attacks on fire-fighters occurring during periods of public

Left:
Essex Fire & Rescue Service commissioned this Dennis RS135 water tender by John Dennis Coachbuilders in 1990. Note the roof-mounted monitor.
Roger C. Mardon

Left:
In 1990 Dennis Specialist Vehicles worked with John Dennis Coachbuilders to produce the Dennis Rapier. A choice of Cummins 8.3-litre diesels was offered for the power unit: either the 240bhp after-cooled 6CTA or the 211bhp 6CT. The body panels were formed from Metawall, a composite material that was strong enough to replace traditional plywood panels and light enough to be built on to the new tubular stainless steel space-frame chassis. The first operational Rapier went into service with Kent Fire Brigade. *Roger C. Mardon*

disorder and, even more sadly and inexplicably, as a matter of routine in some areas. Fire engines have seen the addition of door locks and, occasionally, mesh screens over the windows for crew protection. Now at least one brigade has had to resort to mesh over rear access ladders to prevent children jumping on to moving vehicles. Equipment and even whole vehicles have been stolen. Today's fire engine is therefore likely to have impact-resistant glass and central locking for the cab and all lockers.

In 1990 Halton Fire Engineering Ltd of Widnes announced a new range of appliances using modular construction techniques on any make of chassis. Norfolk Fire Service commissioned 16 water tender ladders on the Mercedes-Benz 1222F chassis over the next two years, and Wiltshire also took four on the 1124AF chassis.

Mountain Range became part of Powersafe Industries Ltd in 1991 and its range of appliances was built by Reliance Mercury Ltd of Halifax. Buckinghamshire Fire & Rescue Service took delivery the following year of a Reliance Mercury rescue pump

built on the Scania G93ML 4x2 chassis. The appliance was fitted with a 400-gallon water tank and carried rescue equipment for road accidents and other emergencies.

In 1992 Carmichael called in the receivers. Carmichael Fire was acquired by Trinity Holdings, owners of Dennis Specialist Vehicles, and its Iveco-Magirus agency went to Powered Access Ltd of Nottinghamshire. Carmichael's Gregory Mill site in Worcester was to close and production would continue at a new Weir Lane factory. Perren Fire Protection Ltd was acquired by Imperial Fire Defences Ltd of Maidenhead.

Volvo's FS7 chassis was introduced in 1992 with a 7-litre 230bhp engine, or an optional 260bhp unit, both offering greater power than the FL6. The first machine was Saxon-bodied and delivered to South Glamorgan Fire & Rescue Service.

In 1994 came the first order for MAN L2000-chassised pumping appliances in Britain, to be bodied by Saxon Sanbec for Devon Fire & Rescue Service. Eighteen were delivered in 1995 on the 10-tonne gvw 10.224F chassis with 220bhp Euro 2 engine.

Right:
Wiltshire Fire Brigade took this Mercedes-Benz 1124AF/Reliance Mercury pump into service in 1992, and it is seen here at Marlborough.
Roger C. Mardon

Left:
Riversdale Fire Defence Ltd (Hughes) was active in appliance building in Ireland. Unusual in having a Godiva front-mounted 1,600l/min pump, this 1993 Hughes-built Mercedes-Benz 814 pump is seen on the run at Ardmore in Ireland. *Gordon Baker*

Right:
In 1993 Dennis Specialist Vehicles announced a new Rapier cab offering more headroom, legroom and low-entry access. This Carmichael-built 1995 Dennis Rapier is pictured while on the run with Durham Fire & Rescue Brigade as a rescue water tender ladder at Bishop Auckland.
Roger C. Mardon

A consortium of 13 brigades drew up a specification for a standard water tender in 1995. Lower purchasing costs and reduced whole-life operating costs, together with high build quality, were the claimed advantages for a purchasing consortium. Crew safety featured in the requirements for cab design, and equipment stowage was flexible to meet the needs of individual brigades. Sliding trays and mechanisms were installed for easy access to heavy equipment. The standard pump was an alloy Godiva 2500, but variations could be selected at extra cost. An 1,818-litre water tank was provided and the norm provided for a 13.5m or 15m ladder to be carried, and one beam gantry for both a 7m and a roof ladder. The final product was the Dennis Sabre bodied by John Dennis Coachbuilders, the first of which was delivered to Wiltshire.

ERF Ltd reappeared on the fire engine scene in 1995 with an EC8.24 WT2-chassised rescue pump built by Saxon Sanbec for Buckinghamshire.

The Dennis XL cab was developed for the fire service in the Netherlands, which wanted a cab for nine crew members. The stretched cab would actually accommodate 10, and the first for the UK was built in 1996 for West Midlands Fire Service, where it was intended to maintain the usual crew of six and use the extra space for in-cab stowage of rescue and casualty trauma equipment. In 1997 Hertfordshire took delivery of Dennis Sabre ML water tender ladders with a medium cab extension into the rear bodywork, which made no difference to the outward appearance of the vehicle, and one XL model on a 4.2m wheelbase chassis with the larger 400mm cab extension.

Left:
The Dennis Sabre was introduced in 1995 as the successor to the RS and SS range of appliances. The first went on the run with Wiltshire Fire Brigade, but this 1998 model is seen in service as a water tender ladder/rescue with Dorset Fire & Rescue Service at Ferndown. *Roger C. Mardon*

Left:
New Volvo FL6.14/Excalibur CBK water tender ladder rescue appliances for Tyne & Wear in 1996 included horizontal and vertical sliding trays to hold the heavier equipment. Rear roof access ladders were eliminated to prevent children climbing on to moving vehicles unseen, and access was relocated to the nearside behind the crew cab.
Roger C. Mardon

The Saxon Volumax cab first appeared at the Fire 96 exhibition. This provided increased height in the rear crew compartment with stowage over the area occupied by the driver and officer.

Right:

In 1997 the Highlands & Islands Fire Brigade, still with a workshop able to undertake appliance bodybuilding, constructed this pump on the Japanese Isuzu 6.2-tonne NPR chassis. Powered by a 3.9-litre four-cylinder turbocharged diesel engine developing 120bhp, the appliance carried 200 gallons of water and retained the familiar arrangement of three lockers each side. It was followed in 1999 by a Mitsubishi. *Geraint Roberts*

In 1998 West Yorkshire adopted the Volvo FL6.14 with 4m wheelbase, rather than the more usual 3.8m, for a prototype appliance designated as a heavy rescue pump. This appliance was the last to be fully built in

Below:

Netherlands-based Plastisol had been producing injected resin bodies for airport crash tenders for some time when Lancashire Fire Brigade approached the company about the production of a more modest pumping appliance. The body and water tank were produced as one integral unit by a process not involving the use of a mould, and the design included a tumblehome to lend styling to the finished vehicle. This method of production offered a significant

weight saving and increase in locker space. Chassis engineering specialist TVAC Ltd of Leyland, Lancashire, was engaged to build the prototype appliance, and its own first fire engine, using the Plastisol body. The Leyland DAF 55-230Ti chassis and steel cab were used, downrated from 14 tonnes to 12 tonnes, with the resin crew cab and fire-fighting body mounted behind. Production models adopted by the brigade differed little from the 1998 prototype seen here. *Leyland DAF Trucks*

brigade workshops, and it provided the basis for future builds by Emergency One (UK) offering increased accommodation for casualty care, rescue and hazardous materials equipment. The 13.5m ladder was mounted on an AS beam gantry, and eight lengths of flaked hose were carried in a locker above the pump bay.

In 1999 the Jaguar car plant at Castle Bromwich took delivery of the company's first Iveco-Ford Cargo 80E15D-chassised fire engine, with bodywork and engineering by Angloco. The appliance carried 800 litres of water and 100 litres of foam, delivered by a Rosenbauer NH30 pump. A lighting mast was installed and a 10m ladder was carried on the roof. In the same year Avon Fire Brigade put the first of 20 new pumps on the run after successful trials of the chassis over the previous 12 months. The machines were built by Saxon on the MAN M2000 14.264F chassis with Allison MD 3060 automatic transmission.

Left:
Browns Coachworks Ltd of Lisburn built this water tender ladder on a Volvo FL6.14 chassis for Dundalk Fire Brigade in 1998. *Gordon Baker*

Below:
In 1998 Grampian Fire Brigade, after using fluorescent paint for over 25 years, adopted a white livery with yellow reflective stripes for its appliances, as demonstrated by this Scania 94D/Emergency One (UK) water tender ladder. *Colin Dunford*

Right:
Gloucestershire chose the MAN 10.224 for this compact water tender ladder built by John Dennis Coachbuilders in 1999.
Roger C. Mardon

Left:
The Mercedes–Benz Atego 1328F fire chassis was launched in 2000 and adopted by Cornwall Fire Brigade with a steel crew cab by S. MacNeillie & Son Ltd of Walsall and remaining bodywork and fire engineering by Carmichael. The water tender ladder shown is pictured at Launceston. Gloucestershire also placed an early order for this chassis and went to John Dennis Coachbuilders for the engineering and bodywork.
Gary Chapman

Right:
In 2000 Dennis Fire announced the Dagger appliance in conjunction with John Dennis Coachbuilders as a response to the growing demand for a compact fire engine of lower cost than a standard water tender. The vehicle was 2.3m wide with a gross vehicle weight of 12.5 tonnes, and the first customers were Devon, with an order for one, and Hertfordshire, with an order for two. *Dennis Fire*

3. AERIAL APPLIANCES

Turntable Ladders

Turntable ladders are used for effecting rescues from high buildings, but are more often employed as water towers from which a powerful jet of water can be directed downwards on to the fire from a water cannon, or monitor, at the head of the ladder. It follows that the ladder must be self-supporting, and while early ladder designs achieved this criterion, they could not be rotated and could only be elevated to one angle. The turntable ladder was developed to remedy these limitations and in recognition of the fact that the increasing height of buildings was putting the upper storeys beyond the reach of existing escapes.

A turntable ladder consists of an extending ladder mounted on a turntable enabling it to be rotated — or trained, in fire service parlance — through 360°. The ladder itself comprises a main ladder fixed to a swinging frame, the angle of which can be varied to elevate or raise the ladder to the required degree of pitch. Within the main ladder is a number of sliding sections which can be telescoped out, or extended, to increase the length of the ladder. Early ladders were made of wood and were operated manually. Operation by gas and compressed air followed in 1901 and 1902, to be superseded by electric and petrol-engined power, leading ultimately to the modern diesel-powered appliance. British appliances have always had the ladder turret mounted at the rear of the vehicle over the back axle.

According to Magirus-Deutz of Germany, to whom many early developments in turntable ladder design are attributed, while a few turntable ladders had been around since about 1800, little in the way of development took place until 1877/8. The first practical ladders were supplied by Stahl and Hönig, and in 1892 Magirus produced an 82ft hand-operated turntable ladder on a four-wheel horse-drawn carriage.

Edinburgh Fire Brigade built its own horsed 65ft turntable ladder in the workshops because there was nothing else available to suit its needs. The manually operated ladder was in three sections and the turntable was made from the fore-carriage of an old fire engine. The apparatus lasted 20 years and it was said to have been the first and only turntable escape in Great Britain for many years. Quite when it was made now seems to be uncertain, but a second 75ft version was built in 1900. Shand Mason & Co was nevertheless offering a horse-drawn four-wheel turntable ladder in 1896.

Manchester and Leicester both had horsed Magirus 82ft gas-powered turntable ladders manufactured under licence by John Morris & Sons of Salford in 1904. Two

Left:
In 1903 Simonis of London introduced to Sheffield the first powered turntable ladder in the country. This was a Braun 75ft ladder powered by carbonic acid gas (carbon dioxide) stored in four cylinders. When the ladders were fully extended, the engine would be thrown out of gear, leaving the gas to escape from a safety valve on the feed pipe. An indicator was provided on the foot of the ladder showing the height to which it was raised, and there was an oil brake for lowering the ladders. Plumbing gear was incorporated enabling the ladders to be kept perpendicular when used on a slope. The whole was mounted on a four-wheeled horse-drawn carriage. *The British Commercial Vehicle Museum Archive*

Right:
Recognising that the ladders could still give good service, in 1921 Sheffield Fire Brigade sent its 18-year-old turntable ladder to Leyland Motors to be reconditioned and mounted on a petrol chassis. The result was a machine that met the standards of a new appliance in terms of road performance and ladder operation. It remained in service until 1928. *The British Commercial Vehicle Museum Archive*

years later Merryweather & Sons supplied Shanghai Fire Department with the first turntable ladder operated by the power of the road engine. The company also offered ladders with independent power units on the same motor chassis as Tottenham's chemical escape tender. Ladders between 60 and 120ft were available, powered by carbon dioxide gas, compressed air or a separate petrol engine mounted on the turntable, with hand-operation as an option for ladders up to 80ft. Simonis relied on petrol-electric traction for a gas-operated 85ft turntable ladder supplied to Glasgow in 1906, in which two electric motors driving the front wheels through a gear were supplied with current from a 30-40hp Mercedes engine.

In 1919 London Fire Brigade had in service four Cedes battery-electric ladders and two 84ft ladders mounted on Tilling-Stevens petrol-electric chassis. Over the next few years the brigade had its existing Magirus ladders rechassised on to the Tilling-Stevens chassis, paving the way for London's last horsed-drawn appliance, a turntable ladder, to be withdrawn in 1921.

In 1920 Tilling-Stevens had petrol-electric turntable ladders in build for Newcastle and Nottingham. Merryweather was advertising its motor turntable ladder in 1921, the engine of which propelled the machine and was used to raise and the extend the ladders. Elevation and extension could be carried out simultaneously. Rotation of the turntable was by hand-operated worm-and-wheel gearing. Machines were delivered firstly to Bristol and Edinburgh, then Norwich the year after. By the following year all movements were mechanised.

Glasgow Fire Brigade took delivery in 1922 of the first Morris-Magirus motor turntable ladder the raising, extending and rotation of which were obtained by direct drive from the Magirus petrol-driven road engine. All three movements of this 85ft ladder could be performed simultaneously. Also introduced at this time was a new rescue apparatus comprising a lifeline and pulley arrangement for rescued persons to be lowered to the ground from the head of the ladder in a sling. This device, said to be the invention of Chief Officer William Waddell of Glasgow, was to become a standard feature of turntable ladders.

Right:
Leicester acquired an 85ft Morris-Magirus turntable ladder on a Belsize chassis in 1913 and based it at the central fire station in Rutland Street. According to the manufacturers this was the first Morris-Magirus to be supplied for the UK market with a motorised chassis, although the ladder was still separately gas-powered. In 1935 the turntable and main ladder were reconstructed in the brigade workshops on a Bedford chassis for future use by the street lighting department.
Malc Tovey

In 1924 Leyland Motors introduced the German-built Metz 85ft turntable ladder. Manchester Fire Brigade had the first of these appliances, which were only available fitted to a Leyland chassis, and 21 years later the ladder set was rechassised on to a 1928 Leyland from Dundee, the wooden ladders of which had broken.

Realising that a turntable ladder could not be used as a water tower without the aid of a motor pump, in 1927 Leicester Fire Brigade arranged with Merryweather for a 250gpm pump to be installed on the chassis of its new 85ft appliance. The novelty of this announcement was somewhat destroyed by the appearance in the same issue of *Fire* magazine of a Morris-Magirus fitted with a rear-mounted centrifugal pump. It was not a new idea on the continent, and Budapest had three Magirus turntable escapes fitted with pumps as far back as 1914.

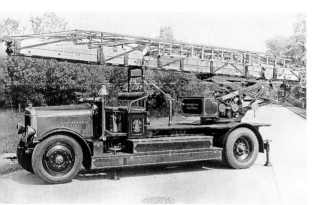

Left:
In 1928 Chief Officer Johnson of West Ham flabbergasted colleagues by putting a motor turntable ladder on pneumatic tyres, this being a time when many commentators believed such tyres were unsuitable for fire engines, and turntable ladders in particular. But by the end of 1933 it was well used, had not suffered a single puncture, and none of the tyres had developed a set, or flat spot. The appliance was an 85ft Magirus turntable ladder pump on a Dennis 5/6-ton chassis fitted with a 400gpm centrifugal pump and a platform monitor. A hook ladder is secured to the trussing of the main ladder.
Surrey History Service/Dennis Bros

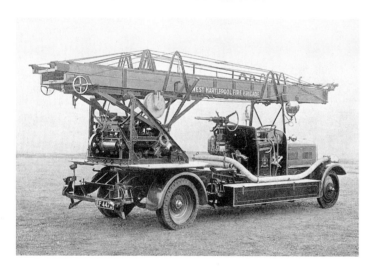

Left:
This Albion/Merryweather 85ft turntable ladder was built for West Hartlepool in 1930 with a 'Hatfield' 275gpm pump. Note the monitor over the pump, which was a common feature of the time. *Author's collection*

Below:
Aberdeen Fire Brigade acquired a new Metz 85ft turntable ladder mounted on a Leyland Lynx chassis in 1930, which was transferred to the North Eastern Area Fire Brigade in 1948. Note the Continental-style detachable hose-reel at the rear. *The British Commercial Vehicle Museum Archive*

In 1932 the first all-steel ladders for the UK were put into service in London and Belfast. These Magirus ladders supplied by John Morris & Sons of Salford offered greater rigidity and extension to 100ft as opposed to the 85 or 90ft of even the longest wooden ladders. They could be allowed to rest against a building and could be used as water towers at maximum extension, something not possible with a wooden ladder. Merryweather's first steel ladder for service in the United Kingdom was delivered to Ilford in 1933 and cost just over £3,000. The 85ft four-section ladder was mounted on an Albion chassis with a 400gpm centrifugal pump, while a separate first-aid pump and 60-gallon tank provided a jet from the 160ft hose-reel for 9 minutes before recourse to an outside water supply became necessary.

The standard 1933 Leyland/Metz turntable ladder was 85ft and made from wood, but steel or alloy ladders were offered in lengths of 92ft, 98ft 6in, 110ft, 131ft and 147ft 6in. Those up to 98ft 6in were mounted on a TLM chassis similar to the FT1 pump, but the longer ladders required a strengthened chassis. A pump could be incorporated, and the appliance supplied to Nottingham was fitted with a 700gpm pump. Orders from Coventry and Portsmouth were for all-steel Metz ladders.

By 1933 Morris-Magirus motor turntable ladders were in service with 17 fire brigades, as well as in London, which had 14. Options included a 400gpm pump, a platform monitor in addition to the monitor at the head of the ladder, rescue line life-saving apparatus, loud-speaking telephone communication between the operator and the man at the head of the ladder, and a searchlight. Merryweather turntable ladders, fitted with pumps, were in service with eight brigades. The Riverdale Committee in 1936 established by a survey of the 770 English and Welsh brigades that 78 turntable ladders were in service.

By 1938 Metz turntable ladders in Germany were fitted with a lift cage that ran on the underside of the ladders to lower rescued and injured persons from upper floors to the ground. In that year Blackpool ordered a 120ft Leyland/Metz turntable ladder with a safety cage for two people, which could be operated at any ladder elevation or extension. A folding ramp was to be provided for entrance from a window or roof, and a short ladder enabled exit once the cage had been fully lowered. However, the war intervened and this appliance was never delivered.

Left:

In 1937 Newcastle City Fire Brigade commissioned one of only two Leyland TLM/Metz turntable ladders to be built with a cab. This five-section ladder set extended to 116ft and the other, which went to Morecambe & Heysham the following year, was a four-section ladder extending to 101ft. *The British Commercial Vehicle Museum Archive*

By 1939 Merryweather's all-steel turntable ladders were in service with or ordered by 26 brigades, including Paris, when Leyton ordered an 85ft model with pump. There were 20 wooden and 32 steel Leyland/Metz ladders. The outbreak of World War 2 put an immediate stop to the import from Germany of Magirus and Metz turntable ladders, and in 1940 the Home Office assisted certain urban brigades with the acquisition of Merryweather 100ft steel turntable ladders. The first Home Office acquisition went to London.

In 1943 the Home Office introduced a three-section 60ft manually operated turntable ladder, which it saw as a useful step between the 50ft wheeled escape and the 100ft appliance. It cost just over a third of the full-size machine and was able to satisfy the requirements of many towns. The basis of the new appliance, built by Merryweather & Sons, was an Austin K4 chassis with a standard two-man cab for driver and officer, and a rear-facing four-man cab or shelter for the rest of the crew. The prototype ladder was made of wood, but the production models were fitted with steel ladders. Generally speaking this appliance was designed for use by crews who were not necessarily qualified as turntable ladder operators, and was generally within safe working limits at any angle of elevation. As the ground clearance at the heel of a fully elevated ladder was only 12in, the vehicle chassis prevented training through a full circle, the maximum radius being about 270°. Fifty of these appliances were built between 1943 and 1944. Many were fitted after the war with the American front-mounted Barton 300gpm pump and continued to give service until the late 1960s or even early '70s.

In 1947 Merryweather & Sons announced a new 100ft steel turntable ladder based on a bonneted AEC Regent double-deck bus chassis with an enclosed single crew cab. The first of the new appliances went abroad, and it was 1949 before any saw service in this country. The South Eastern Area Fire Brigade of Scotland acquired a new Merryweather 100ft turntable ladder on a diesel-powered AEC forward-control chassis in 1951. Driver and officer sat in the cab and other crew members could be accommodated in a rearward-looking compartment. The West Riding took delivery in 1951 of the country's only two Dutch-built Geesink 107ft turntable ladders, which were mounted on a Leyland Comet chassis.

The Metz 100ft turntable ladder was first mounted on the new Dennis F14 chassis, powered by a Rolls-Royce B.80 engine, in 1952. Within a year nine city and county brigades had placed orders.

Left:

Fifty Merryweather 100ft turntable ladders were provided from 1940 to 1944 on either the forward-control Leyland TSC Beaver diesel-engined chassis or the normal-control open Dennis chassis. The earliest were supplied to local authority brigades before nationalisation of the fire service, but this Leyland Beaver was delivered to the NFS at Rugby. *The Westbrook collection*

Right:
One of the 50 Austin K4/ Merryweather 60ft manually operated turntable ladders built for the NFS. *The Westbrook collection*

Left:
In 1955 Birmingham and Lancashire fire brigades each commissioned Magirus 100ft turntable ladders on the Bedford SL chassis with what was marketed as the latest design of all-round-vision cab. The appliances were supplied by John Morris & Sons. The States of Jersey Fire Service commissioned an 85ft Morris-Magirus turntable ladder on the Bedford S chassis two years later. *Vauxhall Motors*

Right:
There were two Metz DL37 125ft turntable ladders supplied for service in the UK, this one mounted on a Dennis F21 chassis to the order of Rochdale Fire Brigade in 1957. Note the five-section ladder, which distinguishes it from the more usual 100ft four-section example. The vehicle, powered by a Rolls-Royce B.80 engine and fitted with a Dennis 500gpm pump, was absorbed into the Greater Manchester Fire Service in 1974 and taken out of service two years later. The other 125ft ladder went to Belfast Fire Brigade in 1955. *Surrey History Service/Dennis Bros*

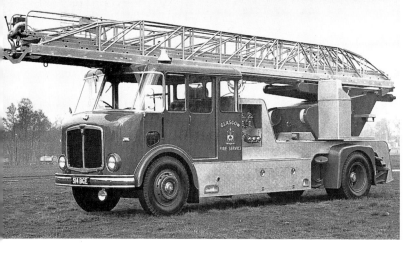

The Leyland Firemaster chassis was used in the construction of two turntable ladders, the first being supplied to Darlington in 1960 by David Haydon Ltd of Birmingham, which had succeeded to the Magirus franchise in 1958. Both were Magirus 100ft hydraulically operated units, and the second went to Wolverhampton in the same year. By July 1962 17 UK brigades had chosen Magirus turntable ladders since the war, and by October 1963 Merryweather hydraulic turntable ladders had been chosen by 50 UK brigades.

In 1967 Merryweather announced the availability of its turntable ladder on the latest AEC Mercury chassis with a 154bhp diesel engine. A specially built version of the Ergomatic cab accommodated driver, officer and a crew of four facing to the rear.

Also in that year, Glasgow Fire Service undertook trials of a detachable two-man cage that could be fitted at the head of a Magirus turntable ladder. A drop-down hinged gate in the front of the cage doubled as an access ladder enabling entry before the ladders were elevated or extended. The brigade was to fit all three of its Magirus ladders with the apparatus. Two years later Glasgow was set to be the first brigade in Britain with the new Magirus LB30 30m turntable ladder, providing a cage operable from the ladder or from the ground; installed at the head of the ladder, the cage was accessible from the ladder or more conveniently from the ground with the ladder depressed below the horizontal.

In 1981 Grampian Fire Brigade commissioned a Metz DLK30 electro-hydraulically powered turntable ladder mounted on one of the last Scania LB81 HS low-level chassis to be built, but the first Scania fire engine to enter service in Britain. The 30m ladder was fitted with a cage at its head from which floodlighting and power tools could be operated, current being supplied by a turntable-mounted 110V generator. It was the first UK appliance to be fitted with a jacking system continuously monitored by a computerised fault-finding diagnostic display panel. This £134,000 machine replaced a Dennis/Metz of 1955 vintage, which had cost £8,265.

The Magirus DLK23-12 turntable ladder mounted on a Magirus low-line chassis toured the UK before being exhibited by Carmichael at Firetech 1982. The appliance had a travelling height of only 9ft 4½in, and featured a variable jacking system and two-man rescue cage at the head of the ladder. It was 1996 before Kent took delivery of the first to see service in this country.

Merryweather introduced the XRL30 turntable ladder in 1982, the whole ladder unit being mounted on a skid with its own three-cylinder diesel engine providing power for the hydraulic operating pump. The unit could thus be mounted on any suitable chassis, and rechassising was forecast as a simpler and cheaper procedure when compared with conventional designs. Fife, Merseyside and Strathclyde fire brigades used this appliance, but it did not have the success of earlier Merryweather ladders. The downfall of Merryweather a few years later left Britain without a turntable ladder manufacturer of its own and brigades were compelled to rely on imports from Germany of the Magirus and Metz ladders, and from France of the Camiva, which Guernsey had first used in 1980.

In 1991 the Metz DLK30 PLC — Programme Logic Control — turntable ladder was exhibited at the Fire 91 exhibition. Fitted with a three-man rescue cage, the 30m ladder was provided with a horizontal/vertical-type jacking system with variable width and electronic ground pressure control. It was the only turntable ladder with a powered lowering capability as well as powered elevation, and could be depressed to 24° below the horizontal.

Right:
In 1985 South Yorkshire Fire Service took delivery from Carmichael of this Dennis F125/Magirus DL30U turntable ladder with a Niveau plumbing turret. The plumbing system automatically operated as soon as the ladder was elevated from the headrest, bringing both the main turret and the ladder set into plumb, and avoiding the need for further plumbing movements during operation. This ideally suited the hilly districts of Sheffield where the ladder was to be used.
Roger C. Mardon

Right:
The Camiva 30m turntable ladder was shown at the Fire '84 exhibition, and in 1985 Dennis Specialist Vehicles announced its availability on Dennis DF and F127 chassis. The ladder was offered with a three-man working platform or three-man suspended cage. By 1987 the demonstration vehicle, on an F127 chassis, had been sold to Merseyside Fire Brigade, which became the first of only five mainland British brigades to order a Camiva ladder. London had three in 1988 at a cost of approximately £210,000 each. *Roger C. Mardon*

Right:
Northamptonshire Fire & Rescue Service chose the Metz DLK30 PLC for its new turntable ladder in 1993, supplied by Angloco and mounted on a Scania 93M-250 chassis. This appliance was stationed at Kettering.
Roger C. Mardon

Left:

In 1997 Magirus introduced the DLK23-12 GL CC articulated turntable ladder, and it was presented in the UK by GB Fire the following year. A 3.5m section at the head of the five-section ladder articulated up to 75° below the horizontal, offering up-and-over access on the same principle as the aerial ladder platform. A three-man detachable cage and a permanent waterway for monitor operation were featured. In 1999 Cambridgeshire Fire & Rescue Service acquired its first of two built on the MAN 18.264 chassis. *Roger C. Mardon*

Left:

In 1999 a Dennis Sabre-chassised turntable ladder was seen for the first time; it was a 1983 Magirus previously with Durham County Fire Brigade, refurbished and rechassised by GB Fire for Dublin Fire Brigade. The chassis was the new Sabre HD of 16.8 tonnes gross vehicle weight with the capacity needed for aerial appliances and other heavy-duty applications. Dublin's new machine featured a double crew-cab. *Dennis Fire*

Hydraulic Platforms

The hydraulic platform consists essentially of two booms with a knuckle joint, which can be elevated and rotated on a turntable. The larger models incorporate a third fly boom. At the end of the upper boom is a cage or platform from which fire-fighters can direct a jet of water from a monitor or effect rescues. One advantage of such an appliance is its ability to take more than one man aloft, or to bring down five or six people at the same time. However, it is a considerably heavier appliance than the turntable ladder, which, contrary to comment elsewhere, it did not replace. Initially, it also lacked the height of a turntable ladder. While some brigades may have found the hydraulic platform entirely suitable for their needs, others found a continuing role for the turntable ladder. Simon Engineering Dudley Ltd had been manufacturing hydraulic platforms for some time before the Simon Snorkel fire-fighting version was introduced at the 1961 Edinburgh IFE/CFOA conference. The demonstrator was a two-boom 65ft model mounted on a Bedford TK chassis. The lifting capacity of the cage was 1,000lb and it could reach maximum height in 50 seconds. As with turntable ladders, some appliances were provided with a built-in pump to supply the monitor at the head. These were known as hydraulic platform pumps, not to be confused with pump hydraulic platforms, described later.

The hydraulic aerial platform was first developed in America where two-boom versions of up to 85ft working height were built. The first British fire-fighting Snorkel was a 65ft model, and before anything bigger could be produced the travelling length had to be reduced from the 55ft acceptable in America to something more compatible with British roads. In 1962 Simon Engineering patented the three-boom 85ft platform with a travelling length of 38ft, the first of which was produced in 1964.

Britain's first operational hydraulic platform was ordered at a cost of £11,000 in 1962 by Monmouthshire Fire Brigade and was delivered in September the following year. Mounted on a Commer 7½-ton chassis, it had a working height of 65ft and a horizontal outreach of 31ft. To protect the occupants of the cage from flame and heat, a water-fog could be engaged to envelop the cage.

Simon introduced the SS85 model with a working height of 85ft in 1964. A short third boom, which folded down in the travelling position, provided the extra height without significantly increasing the overall length of the vehicle. The horizontal outreach was 41ft and the cage would hold five adults. The first orders were from Lincoln and Northern Area fire brigades in 1965, both appliances to be mounted on the new ERF 84RS chassis. The Northern Area Fire Brigade appliance, with bodywork by Fulton & Wylie of Irvine, was delivered in 1967, and was not only the first fire appliance with bodywork built in Scotland but also the first operational 85ft hydraulic platform in Britain.

The availability of Swedish-built Wibe Orbitor hydraulic platforms was announced by Carmichael in 1967. The larger model had a working height of 72ft and a maximum outreach of 35ft 9in at a height of 40ft 7in. The first operational appliances, both 72ft examples, were delivered to Bradford and Belfast fire brigades in 1968, the former on a Bedford KML chassis and the latter on an AEC Mercury.

The 70ft Simon Snorkel was first shown at the IFE/CFOA conference in 1968, when an AEC-chassised model bodied by J. C. Bennett & Co (Coachbuilders) for Glasgow and an ERF/HCB-Angus for Leicester were exhibited. Ireland's first hydraulic platform was a Simon SS70 unit mounted on a Dennis chassis in 1972 for Cork Fire Brigade.

Right:
The Dennis F117 chassis for hydraulic platforms and turntable ladders was exhibited at the 1963 IFE/CFOA conference. A low cab and driving position kept the height of the appliance, fitted with a 65ft Simon Snorkel, to a minimum. The Liverpool Fire Brigade example shown here was powered by a Perkins 6.354 diesel engine, but the Rolls-Royce B.81 petrol engine was an option. *Author's collection*

Bradford's Bedford KML/ Carmichael/Orbitor hydraulic platform went on the run in 1968. This was a 72ft unit with a 1,000gpm monitor at the cage supplied by 3½in pipework along the booms. Compressed-air breathing apparatus could be supplied from a built-in 220cu ft cylinder. Power for the vehicle came from a 7,634cc six-cylinder diesel engine. *Vauxhall Motors*

The three booms of this 85ft Snorkel and the cage incorporating a monitor and lighting are clearly shown in this picture of Manchester's 1971 ERF 84PS/HCB-Angus/Simon SS85 hydraulic platform. The two main booms were powered by dual hydraulic cylinders with hydraulic locking valves, the lower boom traversing up to approximately 84° from the horizontal. The upper boom, with hydraulics connected by two rigid links, had approximately 150° of movement relative to the lower boom. An extension or fly boom, pivoted to the end of the upper boom, folded down to reduce the overall length of the unit for travelling. *HCB-Angus Archives*

In 1972 Simon Engineering introduced a new range of Snorkel hydraulic platforms, the largest, SS289, with a working height of 100ft, soon to become the 103ft SS300. The SS263 proved to be popular with many British brigades and achieved 91ft, with a cage capacity of 800lb and monitor capacity of 500gpm. London's favourite was the smaller SS220 with a working height of 77ft but an increased load capacity of 1,100lb and monitor capacity of 1,000gpm. The new models were marketed alongside the originals for a while, but by October 1975 the Simon SS85 had been discontinued.

The country's biggest hydraulic platform at the time was built in 1978 for Amoco's oil refinery at Milford Haven on a Shelvoke & Drewry 6x4 chassis with a Perkins V-8 640 diesel engine and Allison automatic transmission. The three-boom Simon Snorkel SS300 offered a working height of 103ft with an outreach of 50ft, and could deliver foam on to any of the plant's 70ft tanks.

Right:
Kent Fire Brigade ordered three EPL Firecracker 235 hydraulic platforms for mounting on Dennis F125 chassis, and these were delivered in 1981. They were not satisfactory from the brigade's point of view and were withdrawn from service within a few months, one being returned to the manufacturers for modifications. In 1985 it was concluded that the platforms could not meet the brigade's requirements and they were eventually disposed of without seeing further service.
Roger C. Mardon

Amoco's appliance was equalled in 1985 when Texaco Oil commissioned a Simon SS300 hydraulic platform pump on a Volvo F7.27 6x4 special-purpose chassis with a 6m wheelbase for use at its Pembroke refinery. Fire engineering and bodywork on this appliance, which might well have been the first Volvo in service with a British brigade, were by Saxon Sanbec.

The EPL Firecracker 235 hydraulic platform was exhibited on a DAF chassis at the 1979 IFE/CFOA conference. It differed from the Simon Snorkel in that the upper of its two booms was telescopic, enabling the platform, with a working height of 23.5m (78ft), to present a compact travelling length.

A Bronto 322 hydraulic platform mounted on a Shelvoke & Drewry chassis toured the UK in 1983, but the marque did not catch on until the aerial ladder platform was introduced three years later.

Right:
In 1997 came the last Simon SS263 hydraulic platform to be built, delivered to South Yorkshire Fire & Rescue Service on a Mercedes-Benz 1827 chassis with Saxon bodywork.
Roger C. Mardon

Pump Hydraulic Platforms

A pump hydraulic platform was essentially a pumping appliance with a hydraulic platform, usually of 50ft, installed instead of a wheeled escape or ladder. It thus differed from a hydraulic platform pump (note the subtle distinction), which was essentially an aerial appliance with an in-built booster pump to supply its cage-mounted monitor. The DS50 Simon Snorkel pump hydraulic platform, mounted on a Commer chassis and described as a pump escape platform, was first exhibited at the IFE/CFOA conference of 1962 by Angus Fire Armour. It was fitted with a Coventry Climax UFP 600-700gpm pump, a 90-gallon water tank, 180ft hose-reel and cage monitor. The platform was powered by an independent engine leaving the vehicle engine available to drive the pump through a power take-off.

Worcester City & County Fire Brigade ordered a 50ft pump hydraulic platform in 1963 and this was delivered for service at Kidderminster in April the following year. The Simon Snorkel unit was mounted on a Bedford TK 13ft 11in chassis fitted with a Gwynne high/low-pressure pump, and a water tank of 200 gallons capacity and hose-reels were installed. Such appliances later became popularly known as water tender hydraulic platforms, but few, if any, had the 400-gallon tank of a water tender. Worcestershire's example was fitted with the largest tank for which the manufacturers could find room.

The only 50ft Orbitor was a pump hydraulic platform mounted on a Leyland Mastiff chassis with a 500gpm pump and 165-gallon water tank, put into service at Bradford in 1970, making this the only place in the country to run two Orbitors.

Ireland's first pump hydraulic platform was an ERF delivered to Cork Fire Brigade in 1975, while the following year Drogheda Fire Brigade took delivery of a new Dennis F48. None were ordered for mainland Britain after the mid-1970s until Wiltshire opted for the modern and bigger equivalent in 1993.

Left:
This pump hydraulic platform was ordered in 1970 by Monmouthshire Fire Brigade, the 50ft Simon Snorkel SS50 unit being mounted on a Ford D800 chassis with bodywork and fire engineering by HCB-Angus. A Godiva 500gpm pump was rear-mounted and the appliance carried 250 gallons of water.
HCB-Angus Archives

Left:
In 1993 Wiltshire Fire Brigade acquired this pump hydraulic platform with a 29m Italmec unit mounted on a Scania 6x4 chassis with bodywork by Saxon. The platform achieved an outreach of 15m up to a height of 17.5m. A 2,273l/min pump and 1,000-litre tank were installed, and a top speed of 70mph was attainable with a full crew of six. Two years later the brigade commissioned another pump hydraulic platform, this time an ERF EC10/GB Fire with an Interlift 25m platform built and installed by Camborne Engineering.
Roger C. Mardon

Aerial Ladder Platforms

The 1985 Bronto Skylift range from Finland included the 28.2T1 telescopic platform ladder, which combined a turntable ladder and hydraulic platform in one machine and offered a versatility not previously available. The Bronto's main boom was a telescopic two-section unit of steel box section with the cage attached to a small articulated end boom, giving a working height of 29.5m and an outreach of 18m. A telescopic rescue ladder ran alongside the booms. (On a standard hydraulic platform the ladder, if fitted at all, was designed only as an emergency escape from the cage in the event of a platform failure.) The Bronto Skylift 22.2T1 aerial ladder platform became available on a two-axle chassis in 1987, but the bigger 28.2T1, which required a three-axle chassis, remained dominant in the British market.

In 1987 Kent Fire Brigade became the first to order the Bronto Skylift from Carmichael Fire. Three 28.2T1 machines were ordered, to be built on Scania P92M 6x4 chassis with an 8.5-litre intercooled engine and an all-steel Swedish two-man cab. Cage fittings included an Angus N1800 variable monitor, 110V floodlights and power outlets supplied from a turret-mounted generator, a Paraguard stretcher fitting and Rolgliss rescue apparatus.

The telescopic Simon arrived in 1989, and South Yorkshire Fire Brigade commissioned the first Simon ST300S, a 30m aerial ladder platform built on a Volvo FL10 chassis with Saxon bodywork. An ST240S model was also available, and was acquired by some brigades.

Right:
This 1987 Scania P92M 6x4/ Carmichael/Bronto 28.2T1 was used as a demonstrator before being acquired by Cheshire Fire Brigade.
Roger C. Mardon

Right:
In 1989 London Fire Brigade ordered six Bronto Skylift 33.2T1 aerial ladder platforms from Angloco, which had acquired the UK and Ireland marketing rights the year before. Built on the Volvo F10 chassis with 8x4 drive, the platforms had a working height of 35m and an outreach of 21.5m. They were delivered in 1990 and 1991.
Roger C. Mardon

The next Simon introduction was the ST290S aerial ladder platform in 1991, followed by the ALP 340, launched in 1995. This final unit before Simon ceased trading comprised a telescopic main boom and articulated tip boom, with a working height of 34m and the ability to drop down to 5m below road level. The first appliance was built on a Mercedes-Benz 2631 6x4 chassis and was developed in association with the West Midlands Fire Service. The platform unit and bodywork cost around £250,000 without the chassis, and delivery was quoted as eight weeks from receipt of the chassis. The first operational ALP 340 went into service with West Midlands, which specified a Volvo FL10 chassis with Saxon bodywork.

In 1994 Powys Fire Brigade commissioned a 24m Bronto F24HDT (Heavy Duty Telescopic) aerial ladder platform built by Angloco on a Volvo FL6.18 chassis, but the most popular of the new Bronto range turned out to be the 32m F32HDT, which appeared on Volvo, Scania, ERF and MAN chassis.

Five years later, Mid & West Wales Fire Brigade became the first in the UK to commission a Bronto F32MDT (Medium Duty Telescopic) with a 270kg cage capacity as opposed to the 400kg of the heavy duty unit. Unlike the HDT, which requires a three-axle chassis, the medium duty unit can be mounted on a two-axle vehicle, and the Welsh brigade chose the Volvo FL6.18.

Left:
This Scania 113H–310/Saxon/ Simon ST300S went into service with Essex Fire & Rescue Service in 1992.
Roger C. Mardon

Left:
Durham County Fire & Rescue Brigade put this Volvo FL10/ Angloco/Bronto Skylift F32HDT on the run at Darlington in 1996.
Roger C. Mardon

Remote-controlled Monitors

A class of aerial appliance that found favour from 1969 was the remote-controlled monitor, originally taking two forms, both of which were out of production by 1975. Glasgow Fire Service put its first 'scooshers' on the run in 1969; an entirely new concept in the British fire service, these consisted of a remotely controlled monitor at the head of twin hydraulic booms mounted on a Dennis D chassis with a 900gpm pump and 200-gallon water tank. Bodywork was by J. C. Bennett & Co (Coachbuilders) of Glasgow. The Simon articulated booms, with a working height of 31ft 6in, and the Angus Fognozl monitor branch enabled water to be directed remotely into areas of difficult access. A window breaking device and an AFA Infrastat infra-red detector were also fitted at the head of the upper boom, the latter to indicate the presence of fire to the operator below, who would then direct and open the monitor jet. A demountable Lacon 45ft ladder was secured by brackets to the upper boom, and a lightweight hose-reel with 1in tubing was installed over the rear-mounted pump. In 1970 five bigger Mark 2 scooshers were delivered with 45ft Simon booms having a 360° movement, mounted on the Dennis F46A chassis with bodywork again by Bennett. A Dennis No 3 1,000gpm rear-mounted pump and 300-gallon tank were installed, and this time the Lacon 464 (46ft 4in) ladder was fixed to the lower boom and extended by a power winch protected by safety devices. In 1972 Glasgow Fire Service ordered six new scooshers on Dodge K850 chassis to be bodied by Carmichael and delivered the following year, bringing the total of such appliances to 14.

The second type of remote-controlled monitor was the Simon TSM15 Simonitor, which comprised a single 42ft telescopic boom and escape ladder mounted 8ft above the ground on an appliance with fire-fighting bodywork, water tank and fire pump to give a working height of 50ft. The boom was able to carry up to four men during rescue operations, and, unlike Glasgow's scooshers, was not articulated. A 200gpm to 800gpm water monitor, with interchangeable 3,300gpm foam nozzle, was mounted at the head of the boom and remotely controlled from the back of the vehicle.

The modern remote-controlled monitor is the American Snozzle P.50 articulated and telescopic hydraulic boom, first exhibited at the Fire 95 exhibition by GB Fire of Brierley Hill. The same year, 1995, the company converted a Buckinghamshire 1992 Volvo FL6.17 water carrier into a rescue pump fitted with a P.50 Snozzle. The 15m boom had a piercing nozzle originally designed for use in aircraft fire-fighting and rescue, but it was capable of penetrating a building. Lighting and camera systems were installed at the head of the boom.

One of the rare British flirtations with the water tower appliance occurred in 1999 when West Yorkshire Fire & Rescue Service had the booms from a 1987 Simon SS263 platform rechassised by Direct Access Platforms. The new appliance, on a Volvo FL6.18 with Saxon bodywork, now designated a Magirus Simon Z263, had no cage but a monitor and CCTV camera installation with the intention that it should be used as a water tower and for command observation. However, it did not get beyond the training stage before it was re-instated as a hydraulic platform the following year.

Below:
The first Simonitors were a Dennis supplied to Liverpool and a Ford D1616 supplied to Lancashire in 1971; in 1974 Manchester Fire Brigade took delivery of the first to be mounted on an ERF chassis. This Hampshire Fire Brigade Simonitor, mounted on a Dodge K1113 chassis in 1976, was probably the last to enter service with a British brigade. *Roger C. Mardon*

4. RESCUE TENDERS AND SUPPORT UNITS

Emergency Rescue Tenders

Even though fire brigades are without a statutory duty to respond to accidents not involving fire, they have long been equipped to do so. The Fire Brigades Act 1947 empowered the use of fire service equipment for purposes other than fire-fighting, and many brigades, both before and after that, provided an emergency tender to carry equipment not regularly used. A primary role of such vehicles was always to carry breathing apparatus, which did not start to become regularly available on pumping appliances until the mid-1930s. Heavy fire-fighting gear, such as ground monitors and radial branches, might also have been carried on an emergency tender. Other equipment would have been likely to include a generator, searchlights and rescue gear such as oxy-acetylene cutting equipment and lifting jacks. Putting such items on one vehicle was seen as a way of making it more available than would be possible by dividing it between several appliances. Local needs would determine what

else might be carried, but after World War 2 road accident rescue gear became an ever-increasing necessity.

Two immediate needs of a fire-fighter were fundamental to the role of the emergency tender: a supply of breathing air by one means or another, and illumination of the incident scene. The earliest breathing apparatus consisted of a mask fitted over the mouth with tubes leading out to the open air. Next was an air bag carried on the wearer's back and connected to a breathing mask, which gave 2 or 3 minutes' duration in an irrespirable atmosphere by rebreathing the air in the bag. Then came the smoke helmet and jacket into which air was pumped from the open air through a hose. Various respirators or air-filter masks followed in the belief that filtering out smoke and gases would be enough and without realising the dangers of oxygen deficiency. Self-contained oxygen breathing apparatus sets had appeared by the end of the 19th century but were developed slowly and found more favour in the mining industry than in the fire service.

Below:
Probably the first emergency tender to be supplied in Britain was the Merryweather light and air machine supplied to Manchester Fire Brigade in 1904. This comprised a steam engine on a horsed fire engine carriage driving an air pump and a dynamo for generating electricity. The single-cylinder air pump supplied four valves at the side of the appliance from which air hoses were connected to leather smoke helmets worn by the firemen. Air pumped into the helmet kept the wearer's eyes and nose clear

of smoke. Inside the face mask were a telephone receiver and transmitter linked by wires running through the air hose to a switchboard on the appliance to enable communication between the officer in charge and his men. The dynamo was mounted over the fore-carriage and was driven by a belt from the flywheel of the engine. It supplied eight 32-candlepower electric lights connected by 1,200ft of cable carried on six reels. Originally costing about £850, this machine was still in service in 1920.
Author's collection

Right:
London Fire Brigade commissioned this Dennis emergency tender in 1912. Proto oxygen breathing apparatus sets manufactured by Siebe Gorman are being worn by the crew, and the searchlights are supplied from a built-in generator.
Surrey History Service/Dennis Bros

Early lighting equipment was provided by acetylene-burning lamps, which gave a bright light at low cost, and the vaporised paraffin floodlight was popular in rural areas until World War 2. However, by that time electric floodlights and searchlights had come into widespread use, supplied from the appliance battery or from a dynamo or generator. Cluster lights comprising a number of lights arranged around one reflector also became popular.

The rescue equipment of this time commonly included an oxy-acetylene cutting set, which would be capable of cutting through metal up to 6in thick. Lifting jacks were employed to raise heavy objects and were readily available with a capacity of 20 tons. Electric tools such as drills and saws would be powered by the appliance generator and there would be a range of portable lifting gear including chains and pulleys.

Protective clothing for the crew would have included high-voltage gloves, thigh boots and ammonia masks and suits. The rescue of workmen overcome by gas in sewers was a regular feature of the fireman's job, and a sewer trolley, which could be adjusted to fit various sewer designs for use in removing casualties, was often carried on an emergency tender. Resuscitation apparatus of different types was in use, ranging from a rocking stretcher, which was a mechanical means of applying artificial respiration, to the more sophisticated oxygen or oxygen/carbon dioxide mixture administered through a mask and controlled by the patient's breathing effort.

Below:
In November 1933 Dennis Bros presented its new rescue tender for West Ham at the Commercial Motor Transport Exhibition at London's Olympia, and was awarded the silver cup for the best special coachwork of any vehicle there. The pale blue coach-like appliance was powered by a four-cylinder 80hp engine, and the four-door body provided seats at the front for driver and officer with a seat behind for at least five more men. As well as accommodating five rescue men and their breathing apparatus, the rear of the body provided a fully equipped canteen. Two floodlights were powered by current from a built-in dynamo. Cutting and lifting gear was carried together with a resuscitator, stretcher and medical first-aid outfit. *Surrey History Service/Dennis Bros*

Left:

In 1938 Blackpool Fire Brigade commissioned a 70mph Leyland FK6 multi-purpose appliance, designated an emergency tender, with a rear-mounted 400–700gpm pump. The limousine bodywork by H. V. Burlingham Ltd of Blackpool, open at the back, housed a 20ft extension ladder, 3½kW dynamo, and equipment for fire-fighting, rescue, cutting and lighting, salvage, and communications. Note the individual chrome letters making up the brigade name on the side. *The British Commercial Vehicle Museum Archive*

Postwar specifications for emergency tenders distinguished between those intended for use in large cities and ports, type A, and those intended for smaller towns and ports, type B. The type A tender required enclosed accommodation for a crew of six or eight and equipment to deal with large or difficult fires where breathing apparatus, special equipment or lighting was needed. A PTO-driven generator of at least 5kW, and preferably 7½kW, to power electric tools was specified. The type B tender was less comprehensively equipped and a built-in generator was not required, although a portable unit was usually carried.

Epco Ltd of Leeds launched its Flexi-Force hydraulic rescue kit in 1956, and use of this gear became widespread in the British fire service for many years. The kit comprised a manually operated hydraulic pump that operated various tools via a high-pressure hose connection. Rams of 6 and 8 tons power capacity could also be used as vertical-lift jacks, and other components included a wedge, spreader and various extension tubes. The successive addition of extensions

enabled greater gaps to be created. The kit cost a fraction less than £50 at the time of its introduction.

Emergency tenders varied in size, but in some of the bigger town brigades had developed into large vehicles that did not readily lend themselves to speedy movement through congested urban streets. They were strategically sited but few in number, and as traffic increased dramatically in the 1960s and after, it became more difficult for large emergency tenders to respond quickly. Hence lighter and faster rescue tenders were developed.

The advent of motorways led to new rescue tenders being commissioned by brigades with the latest high-speed roads in their areas, and the Land Rover went on to become popular as a rescue tender in both normal and forward-control versions. From its introduction by Carmichael in 1970, the three-axle Range Rover conversion was used by some brigades, while the Ford Transit, which had been announced in 1965, was used by others. The larger Ford A series also found favour after its introduction in 1973.

Left:

In 1953 Croydon Fire Brigade commissioned a new type A emergency tender built by Hampshire Car Bodies on a Bedford S 5/7-ton chassis. A 7½kW 100V dc generator was driven by power take-off to supply current to a 9in circular saw, Kango hammer, heavy-duty drill and cluster lights. Heavier and frequently used equipment was stowed in outside lockers for ease of access, and interior fittings accommodated other gear in tailor-made compartments and shelving. Compressed-air breathing apparatus was also carried, and the inclusion of salvage equipment enabled the brigade to discontinue the use of its salvage tender. *HCB-Angus Archives*

Derbyshire Fire Service commissioned an emergency tender from the Merryweather Marquis range, built in 1957 on an AEC diesel-engined chassis. Behind the driving compartment was a central electrical compartment housing a 7½kW 110v dc generator to supply the usual range of lights and electric tools. A rear compartment accommodated four crewmen, along with rescue gear that included an Epco hydraulic lifting and spreading set. A Hathaway portable pump and ladders were also carried, and a 12,000lb winch was driven from a power take-off.
Author's collection

Left:

Scotland's Central Area Fire Brigade opted for the Bedford TK chassis in 1962, and this emergency tender was bodied by Metro-Cammell. *Vauxhall Motors*

Bedfordshire Fire Brigade introduced the air lifting bag to the rescue scene in 1969. The 3ft-diameter neoprene bag closed down to less than 4in, but when inflated by compressed air from a BA cylinder, or by vehicle exhaust gases, expanded in 60 seconds to lift a lorry or force apart the sides of a collapsed trench. At the time air bags were not in commercial production and the Bedfordshire brigade developed them from the original design of a Government agency.

From 1972, when Lukas of Germany introduced the world's first hydraulic cutter, powerful tools such as the Hurst 'Jaws of Life', and rescue sets from Lukas and Holmatro, became available. With their ability to cut through and spread apart impacted metal by hydraulic pressure applied to jaws and rams, these were to prove invaluable in road accident rescue and are now widely adopted by the British fire service. The combination tool, which cuts by the pincer-like action of the jaws or spreads by the opening of the same jaws, is often found on fire appliances.

By 1970 nearly a quarter of all calls answered by the fire service were for special services not involving fire, with road accidents making up a high proportion of the number. While still acting under permissive powers, brigades developed a range of special appliances, large and small, for dealing with these eventualities; the rescue role of the fire service was not in dispute even if it was not reflected in the name of any brigade at that time. Indeed, the chief officer of Denbighshire & Montgomeryshire made at that time what was probably the first call for the fire brigade to become the fire and rescue service.

Left:
By 1974 the need for small fast rescue tenders was established, and West Sussex Fire Brigade opted for the 6x4 Range Rover by Carmichael. This appliance was one of two acquired that year with a telescopic lighting mast installed and finished in fluorescent paint. *Roger C. Mardon*

Right:
Well known in Germany for its lighting equipment, Polyma's entry into the British fire engine market came with a successful tender by Polyma UK Ltd for an emergency tender required by Norfolk Fire Service. It was built in 1986 on a Mercedes-Benz 1222F chassis. *Roger C. Mardon*

Left:
This Mercedes–Benz 1124AF/Locomotors rescue support vehicle was put on the run at Newbury in 1993 by the Royal Berkshire Fire & Rescue Service. Designed primarily to deal with road accidents, the appliance is equipped with a rear-mounted Palfinger crane and carries a comprehensive range of rescue tools.
Roger C. Mardon

In 1995 Avon Fire Brigade put on the run a Renault G300.26D vehicle with curtain-side bodywork by A. G. Bracey. Designated a road/rail rescue unit, the appliance was fitted with rail guidance gear enabling it to transfer from road to rail for a fast response to incidents in the Severn rail tunnel. Owned by Railtrack but operated by brigade personnel, the unit had the capacity to carry 12.5 tonnes of palletised equipment loaded and unloaded with a Moffett Mounty fork-lift truck carried on the rear. Two Alumi rail-karts and trailers were also carried to move fire-fighters and their equipment within the tunnel.

Above:
Hampshire Fire & Rescue Service commissioned a Volvo FL6.14/ Saxon special equipment unit in 2001. This vehicle utilises the 4m wheelbase chassis and is powered by a 230bhp turbocharged diesel engine. It carries the equipment not found on front-line pumping appliances, including a decontamination shower, breathing apparatus main control gear, heavy-duty hydraulic cutting and spreading equipment, high-pressure air lifting bags, gas-tight and chemical protection suits, gas monitoring equipment, radiation equipment, 21in positive pressure ventilation fan, aquavac and Acrow props. *Colin Carter*

BA Tenders/Support Units

Before breathing apparatus was universally carried on pumping appliances, the emergency tender was the appliance that carried BA, and the use of medical oxygen in small cylinders meant that the appliance could provide extended support to crews at a large incident. The advent of compressed-air BA created a need for the recharging of cylinders by a compressor, and the breathing apparatus tender or support unit gained in popularity.

In 1986 Greater Manchester Fire Service was operating a breathing apparatus tender that had been built in-house for the servicing and maintenance of BA sets on the fireground. Twelve sets were available for immediate distribution on arrival and 20 fully charged cylinders were carried. Two 600cu ft compressed-air storage cylinders were mounted below the appliance floor and a 240V electricity supply was obtained either from an external source or from a portable generator carried on the appliance.

Right:
In 1995 West Sussex Fire Brigade commissioned two breathing apparatus support units built by Leicester Carriage Builders on the 17-tonne Iveco Super Cargo chassis with a 5.8-litre engine and automatic transmission. The cab accommodated a crew of two and the appliance would normally respond with a supporting pump. A Hamworthy compressor was able to charge eight BA cylinders in 16 minutes with cylinder handling eased by the installation of a hydraulic lift at each side of the vehicle. Air shelters were carried for use at protracted incidents. Also designed for use as control points at hazardous material incidents, the new appliances were provided with appropriate communication facilities and monitoring equipment for use at radiation incidents.
Roger C. Mardon

Chemical Incident and Decontamination Units

The widespread use and transport of chemicals and other hazardous substances led to the introduction of decontamination or chemical incident units from about the mid-1970s. The decontamination element first provides for the vacuum-cleaning or washing of the protective clothing of affected personnel while they are still wearing it, and then its removal in a 'dirty zone' outside the vehicle. Special decontamination showers are available that are designed to wash off materials from protective clothing. After this process, personnel then enter the vehicle to strip, shower and dress in clean attire before leaving the 'clean zone' by a different door.

Chemical incident units carry gear for the protection of crews tackling an incident, and, while there is no standard, can be expected to have available chemical protection suits, gas-tight suits and probably breathing apparatus. Facilities for accessing chemical data will also be provided so that crews will be aware of the properties of the substances with which they are dealing. Frequently equipment for neutralising and safely removing dangerous substances will be carried, very often now supplemented by environmental protection materials supplied by the Environment Agency.

London Fire Brigade responded to over 300 incidents involving chemicals between February 1973 and December 1975, decontamination procedures being necessary at 20 of those incidents since March 1974. Consequently a chemical incident unit was put on the run in 1975 to attend all major alerts with a trained crew under the guidance of a Greater London Council scientific officer. The unit, based on a Commer 3-ton WalkThru van, carried protective suits, compressed-air BA sets with dust filters, a generator to power vacuum-cleaners for dry decontamination, lighting and radiation monitoring sets.

Hertfordshire Fire Brigade put a Ford A0609/Longwell Green chemical cab incident unit on the run in 1979. The body of the appliance was divided into two compartments, with side and rear screens and canopies offering weather protection and shielding. The front compartment housed small items of equipment and acted as a documentation area for personnel and members of the public exposed to risk. The rear area contained larger items such as portable dams and acted as a storage area for contaminated items to be removed from the scene. An LPG-heated shower and dry decontamination gear, including air-line equipment and a special vacuum-cleaner, were available.

In 1994 Royal Berkshire Fire & Rescue Service commissioned a chemical incident unit built by Leicester Carriage Builders on a MAN 11.190 low-floor coach chassis. Full communications facilities were provided in the front section with main scheme and fireground radio, Cellnet fax and phone, and a notebook computer and printer with access to Chemdata, a database of over 20,000 substances and trade name products. Facilities for BA main control, and the display of fireground resources and hazardous material decontamination arrangements, were provided. The rear section of the vehicle contained specialist equipment including a thermal imaging camera and various meters for determining levels of explosive mixtures or oxygen, for example. Decontamination and shower facilities were available in an air tent, while an air shelter offered changing accommodation. The appliance and equipment cost some £137,000, and another £20,000 worth of pollution control equipment was provided by the National Rivers Authority.

Left:
In 1992 Cambridgeshire Fire & Rescue Service converted a Volvo FL6.17 prime mover into a hazardous substances rescue unit, or chemical incident unit, the conversion work being undertaken by John Dennis Coachbuilders. The crew cab accommodated four BA sets, chemical protection suits and wash-basin with hot and cold water. It also offered space for changing and for conference facilities. The rear of the appliance provided a loading area with an HIAB crane and carried a range of equipment for neutralising and removing hazardous substances.
Roger C. Mardon

Incident Support Units

An operational or incident support unit is a vehicle that carries heavy or specialised equipment that it is not practical to carry on a pumping appliance. This makes it sound just like an emergency tender, which in some brigades it is, but in others it has a wider role. It may perform the functions of several special appliances by carrying equipment for heavy rescue and lighting, breathing apparatus support, chemical incidents, foam-making, salvage, canteen facilities or just about anything else that might conceivably be found on a fire brigade's equipment inventory. This assorted apparatus might be carried together in any combination, or the appliance might be arranged so that equipment stored on pallets or in containers can be quickly selected and loaded as required. The appliance itself could be anything from a GMC K30 4x4 to a 25-tonne 6x4 vehicle, according to its role.

The fork-lift truck is now a familiar sight on the back of fire engines such as large operational support units. The Moffett Mounty with origins in County Monaghan, Ireland, was initially launched in 1986 for the agricultural market, but the potential for loading and unloading trucks by a fork-lift carried on the vehicle was recognised. The Kooi Aap from The Netherlands is another piggy-back fork-lift truck, in service with Royal Berkshire Fire & Rescue Service since 1998 and carried on an ERF EC8/Bracey operational support unit.

Below:
In 1991 Avon Fire Brigade put a Moffett Mounty M2003 fork-lift truck into service with its new Renault G300-24D 6x4/Bence operational support unit. The vehicle carried eight 1,000-litre foam tanks on pallets, and the fork-lift, transported by a mounting on the back of the Renault, enabled foam and equipment to be taken across rough ground if necessary right up to the point of use. The Mounty, with its lift capacity of 2,000kg, could also be used for moving goods away from the scene of a fire and for lifting cars involved in serious accidents. Foam compound in sealed containers has a life of 15 years, but was available only in 5-gallon and 1,000-litre sizes. The former, smaller size was considered too small for efficient use on the fireground and the larger, latter size was too big for manhandling; hence the acquisition of the Moffett Mounty, which could unload the eight containers, sufficient for 70 minutes of foam application, and deliver them to the point of use in about 10 minutes. Also available were pallets of salvage equipment, a fuel pallet with tank and jerry cans for refuelling appliances, and a refreshment pallet to provide canteen facilities at protracted incidents. *Roger C. Mardon*

Above:

In 1994 Lancashire Fire Brigade put five
incident support units on the run as
replacements for six emergency tenders. Pumps
had already been equipped with Holmatro
hydraulic cutting and spreading tools
supplemented by Tirfor hand-operated winches,
and the new units carried equipment with
greater capability. High-capacity shears and
spreaders could be simultaneously powered by
an engine-driven hydraulic pump mounted on a
brigade-designed trolley, and outside lockers
were provided with Ratcliff electro-hydraulic
side lifts for the safe handling of heavy
equipment. The appliances were also intended to
act as command and control units for incidents
of up to eight pumps, so a communications area
offered radio, phone and fax facilities. A
workbench for BA maintenance was also
provided. The appliance was built by Fosters
Commercials on a 10-tonne Leyland-DAF 45
series chassis. *Roger C. Mardon*

5. OTHER SPECIAL APPLIANCES

Salvage Tenders

As far back as 1707, insurance companies employed porters to help remove goods from burning premises. Later they established salvage corps to attend fires, with the fire brigade, and protect goods from water damage and the effects of fire-fighting as much as to rescue them from the fire. The Liverpool Salvage Corps was established first, in 1842, followed by London in 1866 and Glasgow in 1873; all three were disbanded in 1984 after the insurance companies decided that their specialist services could no longer be justified. Outside those major centres, salvage corps were maintained as volunteer bodies in Brighton, Horsham and Harrogate, and as municipal bodies in Tunbridge Wells and Belfast. These survived into the 20th century, Belfast being the last to go when it was absorbed by the National Fire Service in 1941.

Initially horse-drawn salvage traps, and later motor tenders, were employed by these corps to transport men and equipment. In the absence of a salvage corps, the work was, and still is, undertaken by the fire brigade.

After World War 2, the Austin K2 auxiliary towing vehicle was often used as a salvage tender by the salvage corps, and by the fire brigades where there was no corps. New vehicles were commissioned and some fire brigades acquired pump salvage tenders, which, apart from the customary role, could be used to pump out flooded premises or as fire pumps.

Equipment became more sophisticated and, by the demise of the salvage corps, vehicle installations included such things as a 7½kW generator and a Dale Stem-Lite floodlighting mast. Typical portable gear might have included a 3kW portable generator, an air heater, a smoke extraction unit able to move 900cu ft of air per minute, deodoriser, portable pumps, and submersible pumps. Traditional canvas salvage sheets might have numbered 40 or more, and a roll of heavy-duty polythene, squeegees, sawdust, buckets and brooms would have been included in the equipment carried to minimise water damage.

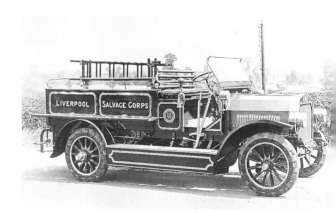

Right:
This is one of two Liverpool Salvage Corps 1911 Dennis salvage tenders, 1911 being the year when the corps acquired its first motor vehicles. The Braidwood body accommodated four men each side, in addition to the driver and officer, and a pair of scaling ladders is mounted above the equipment locker. The Dennis tenders were replaced in 1929 and 1930 and the Braidwood bodies were transferred on to new Chevrolet chassis. *Surrey History Service/Dennis Bros*

Right:
Belfast Salvage Corps acquired this Dennis limousine salvage tender in 1932. *Surrey History Service/Dennis Bros*

Left:
The only Dodge K850 salvage tender of the London Salvage Corps was this HCB-Angus appliance ordered in 1972 and put on the run in 1974. It remained in service until 1982.
HCB-Angus Archives

Left:
In 1995 Hertfordshire Fire & Rescue Service took delivery of two damage control units built by Leicester Carriage Builders on the Iveco-Ford Cargo 100E chassis. Equipment was held in trolleys and included a smoke extraction system and portable lighting, removal being aided by a tail-lift. A lighting mast with sodium lights, instead of the more usual halogen, was installed on the front nearside.
Roger C. Mardon

Foam and Special Media Tenders

Water is generally ineffective against fires in flammable liquids because it sinks below the surface before its cooling powers have time to work. The rise in the use of petrol-engined vehicles, and later in air transport, emphasised the need for a medium that would extinguish burning fuels, and fire-fighting foam was first demonstrated in 1904. This relied on the chemical reaction of the ingredients when mixed together to make a frothy substance the bubbles of which contained carbon dioxide, and it worked by floating on the burning liquid to form a blanket that excluded oxygen and smothered the fire. The Foamite Firefoam Co was established in America in 1918 and within a few years foam extinguishing agents were available in Britain.

A range of products with evocative names appeared: Merryweather's 'Fire Suds', Pyrene's 'Phomene', the General Fire Appliance Co's 'Fire Froth', and 'Fire Snow' by John Morris (Fire Snow) Ltd of Manchester, at pains to point out that it had no connection with any other firm, presumably John Morris & Sons, of Salford.

These were generally available in hand and wheeled extinguishers up to about 34 gallons. In 1922 Foamite Firefoam Ltd of London, recognising the limitations of its wheeled extinguishers, marketed a 220-gallon appliance that produced 1,760 gallons of finished foam and was available for mounting on a four-wheel trailer or on a 3-ton motor chassis.

In 1923 Merryweather announced its 'Fire Suds' foam tender. This was on the company's standard 50hp chain-driven chassis with a Merryweather 'Ravensbourne' pump fitted. The two foam solutions were delivered separately through the two barrels of the pump into two lines of hose and mixed at the delivery branch to make 'Fire Suds' foam. Foamite Firefoam announced its 1,260-gallon Foamite engine for motor chassis or trailer mounting in the same year.

Merryweather built a foam tender in 1924 able to discharge 1,800 gallons of the company's 'Fire Suds' foam. In appearance this was very much like a motor tender of the time, and comprised two tanks holding an alkaline solution and six separate tanks holding an acid solution, all enclosed within the bodywork. A double-chamber rotary pump forced the two liquids through hose lines to a single delivery branch where they were mixed to make foam. Each tank compartment was

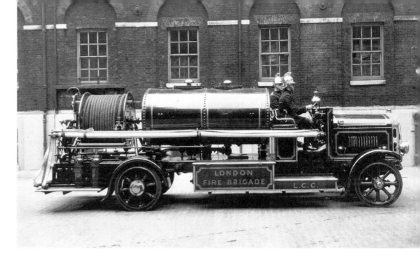

separately controlled by valves and brought into use as necessary. A 50-gallon first-aid water tank and 150ft hose-reel were also provided for use at ordinary fires, together with an overhead extension ladder. This prototype tender was sold to Leeds Fire Brigade in 1925.

In 1924 it was announced that Foamite Firefoam and Tilling-Stevens had co-operated in the building of a 600-gallon motor foam engine: 300 gallons each of acid and alkali solutions were carried in tanks, each discharged at a maximum of 40gpm by rotary pumps to produce up to 700gpm of finished Firefoam through a 300ft hose-reel. The chassis was a Tilling-Stevens 50-60hp petrol electric.

Within a year or two, both Safoam and Henry S. Simonis's portable foam generators had become available, which, even though they did cost about £2 a minute to use, were more convenient than the system of mixing acid and alkaline solutions in great quantity as used on the big foam tender, which they superseded. Then came mechanical foam, made by introducing a compound into the water and mixing it with air to produce foam. The Pyrene foam-making branchpipe was introduced in 1934 and the foam-making

generator in 1948, both requiring a supply of foam compound rather than a dedicated foam tender.

Air-foam pumps had been developed by the early 1930s, and these both mixed and agitated the water and the compound with air, then pumped the finished foam to the fire. This system did need an appliance dedicated to the foam pump, but its use was confined to airfield crash tenders and other specialist applications such as oil refineries.

During World War 2 and in the immediate postwar years, a foam tender was generally little more than a vehicle designed or adapted to carry foam compound and foam-making equipment in larger quantities than could be carried on a pumping appliance. The compound was usually contained in 5-gallon drums and may have been anywhere from, say, the 180 gallons needed to supply one No 10 and two No 2 foam-making branches for half an hour, up to 650 gallons. The use of chemical foam in the fire service had just about died out, although it was still used in fixed installations protecting oil refineries and the like.

From the 1950s foam tenders with foam compound tanks and built-in pumps were becoming popular

among public brigades as well as the specialist industrial brigades. Essex Fire Brigade took delivery of a foam tender built by Sun Engineering (Richmond) on a Dodge 125 6-ton chassis in 1957. This appliance had a Dennis 400-500gpm pump at the rear and water and foam tanks of 350 and 400 gallons respectively. The size and facilities of a foam tender depend on the risks protected by the brigade.

High-expansion foam was introduced in the mid-1960s, a synthetic product that was too light to be projected and was usually fed through large-diameter flexible ducting into confined spaces, such as basements or ships' holds, with the intention of completely filling them. The smaller generators were sometimes carried on pumping appliances, but larger units were normally carried on a foam tender or, in some brigades, on a dedicated hi-ex unit that might have been a small van or trailer.

Left:
This 1981 Ford D1317 multi-role appliance was built by Angloco for the joint brigade of Lindsey Oil Refinery and Conoco at Immingham. The rear-mounted pump was a Godiva of 1,000gpm capacity and the machine carried 900 gallons of water and 200 gallons of aqueous film-forming foam. A 1,000gpm monitor was operated from the nearside platform at the rear of the appliance, which normally carried a 30ft Bayley ladder. High-pressure hose-reels offered a choice of water or foam delivery, and a comprehensive range of fire-fighting and rescue equipment, including air lifting bags, was carried. *Angloco*

Right:
In 1985 the UK Atomic Energy Authority provided two dry powder tenders for protection of the Dounreay fast breeder nuclear reactor. These were based on the Ford Cargo chassis with skid-mounted units carrying two 500kg dry powder vessels and 30m discharge hose-reels. Dry powder is effective for the rapid extinction of certain types of fire, but offers less protection against re-ignition than other extinguishing agents.
Colin Dunford

Airfield Crash Tenders

Civilian Crash Tenders

Early appliances for extinguishing fires in crashed aircraft were at best nothing more than ordinary fire engines modified to produce foam. Often they were light vehicles carrying only portable foam extinguishers and the most basic of breaking-in and cutting tools, and there was no recognisable attempt to produce a purpose-designed crash tender until the 1930s.

In 1933 the recommended apparatus for civil aerodromes was a 30-gallon foam extinguisher mounted on a fast motor vehicle, which should also have been provided with six 2-gallon foam extinguishers to supplement the main unit and two canvas folding buckets. Rescue tools for the extrication of persons from the wreckage were basic and comprised two fire axes, a 6ft-square asbestos blanket with a hole at each corner, two pairs of asbestos gauntlets, and an asbestos helmet. Two salving hooks for adjusting the asbestos blanket (hence the holes) as well as for removing loose aircraft parts, two knives and a pair of bolt croppers were also recommended. A folding stretcher and surgical first-aid kit were to complete the apparatus. For sea aerodromes serving passenger-carrying seaplanes, flying boats or amphibians, a fast motor boat with six methyl-bromide extinguishers for fire-fighting was recommended. For rescue work, two fire axes, a pair of bolt croppers, a grapnel with three prongs and 40ft of chain, two 12ft salving hooks and a knife were proposed. As with the land appliance, a folding stretcher and surgical first-aid kit were to complete the apparatus.

In 1932 John Morris & Sons adopted the 'Suvus' crash tender as a standard type following its successful use at a Manchester air crash. A 30-gallon foam tank was fitted from which a jet could be thrown 70ft, and 12 1-gallon foam extinguishers were carried. Also carried were asbestos blankets, a 16ft preventer hook, and various demolition, cutting and emergency tools. It was claimed that this appliance could travel over rough surfaces in all weathers at 50mph.

A CO_2 tender with nine CO_2 cylinders and three CO_2 hose-reels installed, as well as a 50-gallon water tank, was built in 1933. Two years later Croydon Fire Brigade, protecting an airport with nearly 150 aircraft movements a day, demonstrated the tackling of fire in a crashed aircraft by using a rescue tender and motor pump.

Left:
In 1939 Leyland Motors built this crash tender for Liverpool Fire Brigade to provide cover at the city's Speke Airport. The appliance was built on a 6x4 chassis fitted with a 500–700gpm centrifugal pump and 650-gallon water and foam tank. Four hose-reels with permanently coupled Pyrene FB2 foam branches, each with an output of 450gpm, were provided to give a total foam capacity of 1,800gpm. *The British Commercial Vehicle Museum Archive*

The NFS crash tenders, such as they were, generally comprised a mobile dam unit with foam-making equipment and something like the basic rescue gear already described. Asbestos suits were sometimes carried to allow firemen to effect rescues from the heart of a burning plane, and 2 minutes was considered the maximum safe period for the wearing of this hot and heavy protection. Among various local improvisations, several utility crash tenders were built in 1942 and 1943 by the NFS in Durham and North Yorkshire by mounting a 600-gallon water tank and light pump on a 4/6-ton lorry chassis. Fifty gallons of foam compound in cans and two No 2 foam branches, together with basic rescue gear, enabled these inexpensive machines to provide a simple and effective response to air crash fires. The use of a closed tank offered an advantage over the mobile dam unit in that the tender would arrive at an incident without having spilled a proportion of its water.

In August 1947 the Government assumed responsibility for the fire defences of civilian airports with personnel recruited primarily from the armed forces and ex-members of the NFS. By 1950 state-owned aerodromes had been divided into four classes, and private licensed aerodromes into three, according to the all-up weight of aircraft normally using the facility. Each class was required to have available minimum quantities of foam and CO_2 with minimum discharge rates, and discretion could be exercised as to which appliances were provided to meet those requirements. In 1956 the categories were extended, Grade 1 being the highest and Grade 7 the lowest. Frequency of movement as well as aircraft weight was now taken into account. The categories have changed but the same principle applies today.

In the late 1940s and early 1950s a typical civilian airfield crash tender would carry up to 500 gallons of water and 100 gallons of foam compound in its tanks and would probably have a CO_2 system installed. The 1950 Thornycroft Nubian 4x4 built by the Pyrene Co and commissioned by the Gloster Aircraft Co primarily for the protection of its Meteor jet aircraft is representative of the period. A 500-gallon water tank

and 40 gallons of foam compound enabled a 3,500gpm foam output for 1½ minutes when both of the Pyrene mechanical foam generators built into the front wings were in operation. Each generator had an 80ft hand-line connected. A Coventry Climax pump was installed with 4in suction and two 2⅛in deliveries terminating outside the cab. The CO_2 installation comprised six 50lb cylinders connected to two 100ft hose-reels.

It seems that it was not until the mid-1950s that the roof-mounted foam monitor appeared with any regularity on crash tenders. Jersey Airport took delivery of a new Thornycroft 6x6 built by Sun Engineering (Richmond) Ltd of Surrey in 1955 with the single crew cab surmounted by a monitor for foam delivery. Designs also evolved enabling both water and foam supplies to be replenished from supporting appliances, or some other source, without interfering with foam production. Both Sun Engineering and Pyrene were to play a leading role in crash tender production for over a decade, building appliances for aircraft manufacturers and airlines, as well as for airport fire services. Thornycroft provided the chassis for just about every 6x6 vehicle and many of the 4x4 tenders as well.

The deployment of separate rescue tenders was common at the time, and a vehicle designed for air crash rescue would carry a 12in electric saw or perhaps an air-powered saw, impactor chisels and reciprocating hacksaws, compressed-air BA sets and an extensive range of breaking-in, cutting and lifting gear. Either 200lb or 300lb of dry powder for quick fire knock-down would also have been available. The four-wheel-drive Bedford R was a popular choice for rescue tenders and was adopted by the Ministry Aerodrome Fire Service for this role and for the CO_2 tenders that it then had in service.

AEC launched a heavier version of the Thornycroft Nubian 6x6 crash tender chassis in 1964, with a 20-ton gross vehicle weight and designated the Nubian Major. This went on to provide the basis of many crash tenders and remained unchallenged in its class until Reynolds Boughton introduced its Griffin and Taurus chassis in the 1970s.

Above:
In 1957 the first of a new range of crash tenders for the Ministry of Transport & Civil Aviation Aerodrome Fire Service, formed in 1954, were delivered to London's Heathrow Airport. Three types were built by Sun Engineering on Thornycroft Nubian 6x6 chassis with Rolls-Royce B.81 eight-cylinder petrol engines, and Merryweather 500-700gpm pumps. A foam tender with 800 gallons of water and 100 gallons of foam compound on board would normally be supported by a water tender carrying 960 gallons of water and 35 gallons of foam, but each could operate independently of the other. The third machine was a foam/CO_2 tender with 550 gallons of water, 100 gallons of foam, and 600lb of CO_2. *Author's collection*

Right:
Another weapon in the armoury against aircraft fires was the Cardox low-pressure CO_2 tender. This 1959 unit mounted on a Thornycroft chassis carried 6,000lb of CO_2 kept at a low pressure of 300psi by refrigeration. The electro-hydraulic boom of this appliance, in service at London Heathrow Airport, discharged gas at the rate of 2,500lb per minute. *Author's collection*

Right:
This Bedford R/Miles crash tender went on the run with British Aerospace at Brough in 1961, displaying the typically Miles roof-line and styling.
Andrew Henry

Left:
Crash tenders were getting noticeably bigger by the time the 35-ton Reynolds Boughton Griffin/Pyrene Pathfinder crash tender was first seen as an export order for Yugoslavia in 1971. The machine, fitted with a 13,500gpm foam monitor, carried 3,000 gallons of water and 360 gallons of foam liquid. The first vehicles for British use were ordered by Manchester Airport and delivered in 1973 at a cost of about £70,000 each, and Liverpool followed suit in 1976 with this predominantly yellow appliance, by which time Pyrene had been subsumed by Chubb Fire Security.
Andrew Henry

Airport crash tenders are expected to reach the end of any runway and, if required, to produce foam within 2 minutes of call, subject to a long-stop response time of 3 minutes. Similar response times apply to all other parts of the airport used by planes immediately before and after take-off. In the mid-1970s the rapid intervention vehicle was coming into vogue, a smaller and faster vehicle than a major crash tender designed to effect a quick knock-down or contain fires in crashed aircraft pending the arrival of heavier appliances. This increased the opportunities for rescue and consequently the chances of passenger survival.

Carmichael's new Jetranger crash tenders were introduced in 1976, the 1000 and 2000 models on four-wheel-drive chassis and the 3000 model on a six-wheel-drive chassis. The engine was installed behind the front axle and cab, producing a quieter environment for the crew, while a low-line chassis and tailored water tank gave a low centre of gravity. The Jetranger series was destined to remain until the introduction of the Cobra in the early 1990s.

The rear-engined Scammell Nubian Major 2 was launched in 1977 with 4x4 and 6x6 drive configurations, and marked the end of the Thornycroft marque, which had dominated the crash tender scene since the 1950s. The company had been taken over in 1961 by AEC, itself acquired by Leyland, which already owned Scammell; it is perhaps surprising that the Thornycroft name was not lost earlier. However, that was only the start of a series of changes to affect this specialist market. Leyland was sold to DAF in 1987 and that led to the demise of Scammell the following year. Unipower promptly stepped into the breach, only to become part of Alvis in 1994. Finally, in 2000, Carmichael International acquired the rights to the Unipower fire chassis.

Reynolds Boughton chassis had made their mark and Shelvoke & Drewry were next on the scene in 1977. The Shelvoke & Drewry CX chassis was to be linked with the Carmichael Jetranger until the turn of the decade.

Left:
This Thornycroft Nubian Major was built by Carmichael for the Royal Aircraft Establishment in 1976. The photograph illustrates the BCF installation with the cylinder mounted centrally on top of the body and the hose-reel, which terminates in the applicator held by the fireman, to the offside. *Manston Fire Museum*

Right:
The Gloster Saro Javelin 6x6 crash tender was announced in 1977 and offered a water tank capacity from 8,172 to 10,000 litres and a foam tank capacity from 981 to 1,200 litres. It was first shown at the 1978 Farnborough Air Show on a Reynolds Boughton 6x6 Taurus chassis with a General Motors V12 diesel and automatic transmission. With a tank capacity of 2,250 gallons, the displayed vehicle was able to deliver aspirated foam at 10,000gpm. The first operational Javelin appeared in 1979 at Cardiff Airport, but it would not be until 1980 that the BAA would place an order. *Manston Fire Museum*

Left:
In 1981 Angloco built this crash tender on the new Shelvoke & Drewry CY 4x4 chassis for Humberside Airport. Powered by a Cummins diesel, the appliance accelerated from 0 to 50mph in under 40 seconds and had a top speed of 70mph. An 800-gallon water tank and 100-gallon AFFF tank were installed, with a 100kg BCF unit providing a supplementary fire-fighting medium. The vehicle could pump while stationary or on the move. *Angloco*

Right:
From 1980, for five or six years, the Reynolds Boughton RB44 Apollo 4x4 chassis was used as the basis for a number of emergency tenders and rapid intervention vehicles, with Luton and Plymouth airports being among the first to chose Carmichael-built RIVs. Birmingham Airport commissioned an emergency tender with first-strike fire-fighting ability in 1982, a combination that indicated how the role of the separate airport rescue tender was being subsumed by that of the RIV. This 1983 model by HCB-Angus served at Exeter Airport. *Colin Dunford*

Gloster Saro acquired the fire crash tender business of Chubb Fire Security in 1984, but in 1987 was itself acquired by Simon Engineering for £6.8 million. The new organisation continued to build Protector crash tenders and RIVs, and mostly on its own chassis, but in a little more than 10 years had ceased trading.

Timoney displayed its Mk III fire crash tender on the company's own design of 8x8 chassis with a 540bhp engine at the Fire 87 exhibition. A 13,250-litre water tank and 1,325-litre foam tank were fitted, with 227kg BCF also available. Such vehicles were already in service at Shannon, Cork and Dublin airports, the Mk IV was under development, and the Irish Air Corps had in service Timoney 4x4 rapid intervention vehicles. By 1987 the company had developed an appliance powered by a 380bhp Detroit diesel and able to carry 5,000 litres of water and 600 litres of foam compound.

In 1990 Reynolds Boughton introduced the Barracuda crash tender available as a 4x4 or 6x6 vehicle with a top speed of 75mph. The four-wheel-drive model could carry up to 6,000 litres of water, while the six-wheel-drive version had a water tank capacity of up to 10,000 litres. The following year, Unipower unveiled the Nubian 4x4 and Nubian Major 6x6 crash tender chassis, which were 0.4m wider than their predecessors.

In 1995 the British Airports Authority took delivery of its first new Kronenburg crash tenders designed to replace the Javelin and Meteor appliances then in service. Sixteen MAC-11 32-tonne crash tenders and four MAC-08 rapid intervention vehicles were to be supplied under a contract worth £9 million. The vehicles were built by the Dutch-based company, part of Rosenbauer, which discontinued production in the Netherlands in 2000 and transferred all business to its base in Austria.

Left:
In 1991 Simon Gloster Saro produced a new range of crash tenders, the Protector rapid intervention vehicle carrying up to 6,000 litres of water. The Highlander Protector, with Timoney chassis and independent suspension, was ordered for eight airports of the Highlands & Islands Airports Authority, the one shown being allocated to Sumburgh, Shetland Islands, in 1993. *Colin Dunford*

Left:
The Kronenburg MAC-08 4x4 rapid intervention vehicle is powered by a 710bhp Detroit diesel through Allison automatic transmission, giving it a top speed of 70mph and acceleration from 0 to 50mph in 25 seconds. The integrated glass-fibre tank holds 6,000 litres of water and 840 litres of foam. A Kronenburg I pump of 6,000l/min output is powered by a separate Detroit diesel engine and supplies the 4,500l/min roof monitor and 2,000l/min bumper turret.
Rosenbauer

Right:
This Reynolds Boughton Barracuda Evolution 6x6 crash tender was new to Cardiff Airport in 2000. The 30-tonne vehicle, built on the company's Aquarius chassis powered by a 710bhp Detroit 8V92TA diesel engine, carries 10,000 litres of water and 1,200 litres of foam, and is fitted with a Godiva 5,300l/min pump. Two 35kg BCF trolleys are carried for secondary media. *Colin Dunford*

Military Crash Tenders

As with civilian crash tenders, the earliest military airfield appliances were adapted from standard fire engines and from about 1922 Leyland/Merryweather pumps were fitted with a 30-gallon chemical foam unit behind the seat.

During the 1930s six-wheeled Ford AA and Morris Commercial appliances were in service, with three bromylene 30-gallon Froth foam units mounted over the rear wheels. Another machine of the interwar period up to 1939 was the Ford BB, which carried three banks of two 80lb CO_2 cylinders, each connected to a hose-reel terminating in a discharge horn. A pressurised 40-gallon water tank was also installed, with the forward-most hose-reel connected to it.

Right:
This 30cwt Crossley P-type of about 1930 carried 30-gallon foam cylinders with a methyl-bromide additive that was believed to improve its extinguishing ability. The tall vertical cylinders above the foam containers held 1½ gallons of methyl-bromide, but other models held 1 gallon in containers housed within the foam cylinders. Some versions of this appliance provided stretcher accommodation between the side hampers for the transport of a casualty. Note the hand-rung 'bell' behind the driver's seat.
Manston Fire Museum

Right:
The streamlined tender on the 3-ton Crossley 6x4 military chassis was introduced in 1934, and the one shown here was photographed at RAF Scampton in 1936. A 200-gallon water tank and 20 gallons of foam compound were available for foam production and discharge by twin air-foam pumps driven by a power take-off in the cab. The streamlining was not an exercise in styling, but was designed to facilitate cleansing and decontamination after a gas attack. Note the Gruss air springs on the front, which were connected to the leaf springs. *Manston Fire Museum*

The well-known Fordson WOT1 was introduced in 1942 and resembled the Weeton-type in appearance, but had an open cab with a canvas hood and an elliptical 300-gallon water tank. There was no CO_2 installation on this machine and the air-foam pump was driven by a transfer gear from the road engine. A 1944 variant incorporated a light steel folding tower at the rear which could be raised to attack fires in larger aircraft with the monitor at the head; two further monitors were installed at the base of the tower. The Fordson WOT1A 1945 monitor tender was produced next, and the Austin K6 6x4 3-ton chassis came on the scene in 1946 for use as a carbon dioxide tender. The initial rescue truck of the 1944-5 period was a 5cwt 4x2 Willys Jeep or Ford.

In 1956 the Tecalemit subsidiary, Foamite Ltd, announced an order for dual-purpose fire and crash tenders from the Ministry of Supply. These were to become the Royal Air Force Fire Service DP1 appliances, with a Thornycroft Nubian 4x4 chassis and Rolls-Royce B.80 engine. A Coventry Climax UFP2 500gpm pump was rear-mounted, and a 700-gallon water tank and 35-gallon foam tank were fitted, enabling the appliance to operate as a crash truck or as a domestic truck. The Thornycroft Nubian 6x6 DP2 followed in 1959 and was fitted with a 1,000-gallon water tank and 50-gallon foam tank.

Making its first public appearance at the Farnborough Air Show in 1954 was the unusual six-wheel-drive Alvis Salamander crash tender built by Pyrene. This was developed from the chassis/hull of an armoured fighting vehicle and was powered by a Rolls-Royce B.81 rear-mounted engine. With Pyrene foam-making equipment and chlorobromomethane equipment by the General Fire Appliance Co, it was taken into service by the Royal Air Force and designated the Mk 6 crash tender. Forty were supplied in 1958 with 700-gallon water and 100-gallon foam tanks, and 16 gallons of CBM discharged via hose-reels on each side. A 2,500gpm monitor was roof-mounted over the cab.

Left:
In 1940 the Royal Air Force needed a large number of crash tenders at short notice, and the Air Ministry designed and produced an improvised vehicle based on the Fordson WOT1 3-ton 6x4 chassis with steel cab. This became known as the Weeton-type, after the RAF Fire School at Weeton, near Blackpool, where it was developed, and about 350 were built. *Manston Fire Museum*

Left:
In 1952 the first new crash tenders for the Royal Air Force were deployed. Designated the Mk 5, they were built by Sun Fire Engineering on the Thornycroft 4x4 TF chassis with a 220gpm pump, 400-gallon water tank and 60-gallon foam tank. No monitor was fitted and the maximum output was 2,300gpm of foam through two of the four 60ft foam-lines for 2¼ minutes without replenishing. One 50ft length of water delivery hose with a branchpipe was provided at each side.
Manston Fire Museum

Modifications led to the Mk 6A, two batches of which were supplied in 1960 and 1965. Differences from the Mk 6 included the provision of a split-type power take-off enabling foam production on the move, flaked CBM hoses, and a glass-fibre rather than aluminium foam tank. The Mks 6 and 6A were original builds, and the Mks 6B, 6C and 6D were modifications thereof, the most noticeable of which was the provision of a bigger dual-position 2,500/5,000 gpm monitor. In 1958 and 1959 Foamite also produced crash tenders on the six-wheel Alvis chassis, but they were not used by the British armed forces.

Right:
The Alvis Salamander/Pyrene crash tender depicted here is one of the original Mk 6 versions in its unmodified state. The original 2,500gpm monitor over the cab and the nearside CBM hose-reel are the distinguishing features. The 16-gallon CBM installation was later converted on all variants to a 10-gallon BCF system. *Crown copyright*

Right:
The Land Rover-based ACRT — Airfield Crash Rescue Truck — was built by Foamite from 1957 and, after their merger in 1967, by Merryweather. It was designed to fulfil three main functions: the immediate rescue of persons from crashed aircraft, the extinguishing of aircraft wheel brake fires, and escorting aircraft to dispersals as a precautionary measure against taxying fire hazards. There were a number of variants, but the later models were typically provided with two Foamite 200lb pressurised chemical powder cylinders, CO_2 and dry chemical portable extinguishers for fire-fighting. A 20ft light alloy ladder, searchlight, floodlight, pneumatic circular saw and hand tools were provided for rescue work. *Stuart Brandrick collection*

Right:
The 1971 Thornycroft Nubian 6x6/Gloster Saro DP3 was built in small numbers to support foam crash trucks at aircraft fires or for use as a domestic truck. A Coventry Climax pump, 954-gallon water tank and 55-gallon foam tank were installed. A foam monitor was mounted above the cab and two hand-lines were provided. *Andrew Henry*

The manual gearbox Thornycroft TFA 6x6 Mk 7 and the automatic Mk 7A were further Pyrene productions for the RAF. The first TACR — Truck Airfield Crash Rescue — was delivered in 1970 as a replacement for the ACRT and was originally finished in Weidolux fluorescent red paint. This was an HCB-Angus product, also built on a Land Rover chassis. The main fire-fighting equipment consisted of a 100-gallon tank containing a light water or protein foam mixture, a single-stage pump to pressurise the mixture, and two 100ft hoses for distribution. This marked a change from the dry powder used for the initial attack of aircraft fires. The truck was manned by a crew of three, a driver and first officer seated in the cab and a crewman in a backward-facing seat at the rear of the truck.

The light water (AFFF) of the TACR was to be backed up by fluorinated protein foam instead of standard protein foam. This meant that the primary truck with water and blown foam system for protein foam, and the supporting dual-purpose truck, would be replaced by a Primary 1 Thornycroft Nubian Major/Dennis foam tender with 1,250 gallons of water and 125 gallons of fluoroprotein foam liquid (the Mk 9), and a Primary 2 Bedford R/Pyrene foam tender with 460 gallons of water and 80 gallons of fluoroprotein foam liquid (the Mk 8).

The prototype replacement for the Land Rover

TACR was built in 1975 by Gloster Saro on a Range Rover with a 6x4 trailing rear axle conversion by Carmichael. Designated the TACR2 and carrying 200 gallons of AFFF, deliveries were made in 1977. In 1982 HCB-Angus built 18, and from 1985 the TACR2A was built by Carmichael.

The Scammell Nubian 2 4x4 chassis with rear-mounted Cummins V903 engine and Allison automatic transmission provided the basis of the new Mk 10 crash tender of the Royal Air Force in 1981. Built by Carmichael with a four-man cab, the vehicle featured a central driving position and was designed to be air-transportable. A pre-mix foam tank of 2,700 litres capacity and a Godiva 4,500l/min pump were installed, and the dual-output roof monitor delivered 1,360 or 680l/min. A 100kg BCF unit was also available in the offside rear locker. A number of variants were produced: the Mk 10A had separate water and foam tanks and a rear locker on both sides, the Mk 10B of 1986 had a twin-stage turbo engine distinguished by a raised back end, the Mk 10C was a Royal Navy machine with a high-pressure pump, and the Mk 10D carried hydraulic rescue gear. The Mk 10E was a refurbishment by Simon Gloster Saro of original builds.

Left:
The Queen's Flight Fire Service was established in 1954 to protect the helicopters of the flight on their travels, and in 1985 three of these Bedford TK helicopter support vehicles built by Edgehill of Hook were put into service to replace their 1968 counterparts. Each not only carried 900 litres of AFFF and a range of rescue equipment but also 545 litres of aviation fuel as the crews were responsible for providing ground back-up services for the helicopters when they were away from base. *Colin Dunford*

Left:
Ten Scammell Nubian 4x4/Carmichael Mk 10C crash tenders were built for the Royal Navy in 1986. They were to the same specification as the RAF's Mk 10B, but fitted with a multi-pressure pump to provide a high-pressure fog hose-reel system. This one is seen at RNAS Yeovilton. *Roger C. Mardon*

A replacement for the Mk 9 crash tender was agreed in principle in 1979. The Scammell 6x6/Gloster Saro Mk 11, powered by a Cummins V8-500 diesel, was the end result, with a cab layout and central driving position following that originally designed for the Mk 10. An order for 24 vehicles was submitted and deliveries started in 1984. All appliances carried 1,250 gallons of water, 150 gallons of foam compound, and 100kg of BCF, but eight were equipped with a hydraulic ladder and platform for access to the tail engines of TriStar aircraft in service with the RAF; these were designated Mk 11A. The Royal Navy version was the Mk 12 built by Carmichael.

The Defence Fire Services were established in 1990 by the amalgamation of the Navy Department Fire Prevention Service, the Army Fire Service and the Air Force Department Fire Service. The fire services of the Royal Navy aircraft handlers of Flag Officer Naval Aviation and the fire prevention officers of the Procurement Executive were unaffected by the change. (The Procurement Executive is an MoD organisation charged with testing, researching and purchasing for the Ministry.) The new service had nearly 500 vehicles. The Army fire-fighters were all civilians, while the RAF had both servicemen and civilians.

Above:

A new range of crash tenders came into service in 1998. The Unipower 6x6/Carmichael major foam vehicle was powered by a Detroit diesel giving it a top speed of 66mph and fuel consumption of 4½mpg. A 3,000l/min roof monitor, 1,000l/min bumper monitor and under-bumper hand-lines of 225/450l/min were supplied via a Godiva pump from 1,500-gallon water and 180-gallon foam tanks. Other principal equipment included a high-pressure hose-reel with Akron gun, 240V on-board generator and floodlight mast. The new Alvis/Unipower 4x4 RIV entered service in 1999. *Roger C. Mardon*

Hose Carriers and Layers

When water has to be pumped for a considerable distance, it follows that a considerable quantity of hose will be used — maybe more than is available on the pumping appliances attending an incident. Hose carriers were used to carry hose stowed in rolls, as it was on a pump, but in greater quantity, although the need for them diminished as hose accommodation on motor appliances increased. Four-ton lorries were used by the NFS to carry about 6,000ft of hose, which could be delivered to pumps already at work or along the route of a water relay for laying by hand, but any suitable vehicle or trailer might have been pressed into service.

Right:
Cornwall Fire Brigade commissioned this Drake-bodied Austin FFG as a hose carrier in 1961. Hose-reels and an auxiliary pump were fitted, and in this picture two portable pumps are seen on the rear platform, which was enclosed by the hinged ramp and the folding doors on each side. The appliance was later converted to a foam tender.
Roy Goodey collection

Hose layers are more common in brigades covering substantial rural areas, but their value is also recognised by urban brigades. In the traditional appliance, coupled hose is flaked along the bed of the vehicle so that it can be paid out from the rear over a ramp in single or double lines at speeds of up to 30mph. It was not uncommon for a bicycle to be included in the equipment, certainly into the 1960s, so that a fireman could inspect the lines of hose when laid and keep them clear of obstruction. Hose layers also carry various fittings and accessories, including ramps to enable traffic to cross the lines without causing damage. During the war ramp lorries were employed to cope with the vast amount of hose that crossed city streets.

In 1993 Northamptonshire Fire Brigade had a new hose layer with a motorised drum at the rear for lifting hose during recovery of the 125mm hose used, and that system has now developed into the Angus Fetch hose retrieval system. Hi-Vol 125mm hose is laid from the bed of a hose layer in the normal way. However, recovery is undertaken by driving forward adjacent to the hose line, which is picked up with aid of a hydraulic lifting mechanism and retrieved under hydraulic power for reflaking in the hose bed.

Left:
In 1936 the first purpose-built hose-laying lorry, a Dennis Lancet, was delivered to London Fire Brigade. This carried 84 100ft lengths of 2¾in rubber-lined hose, which could be laid out in up to seven lines simultaneously behind the vehicle as it drove along. The hose was flaked over 14 saddles in the appliance.
London Fire Brigade

Left:
In 1991 Buckinghamshire Fire & Rescue Service took delivery of a four-wheel-drive Iveco-Ford TurboDaily hose layer built by Chambers Engineering of Waddesdon. The appliance carried 1¼ miles of hose and, with a two-man crew, could lay hose across ground often inaccessible to other machines. The brigade operated rapid intervention pumps with a reduced hose capacity and the hose layer was also able to support them. *Charles Keevil*

In 1998 West Sussex Fire Brigade took delivery of a new type of hose laying and retrieval system based on a Mercedes-Benz 2524 6x4 25-tonne chassis with Hillbrow curtain sides. The vehicle carries 1,200m of 150mm lay-flat hose on each of two Hannay hydraulically powered drums of American design, and laid at a speed of up to 10mph. The system, installed by John Dennis Coachbuilders, also provides for powered retrieval of the hose, and closed-circuit television enables the activity to be monitored from the cab. Extra hose and fittings can be carried on pallets, and a Moffett Mounty fork-lift truck for handling is available for mounting at the rear. *Roger C. Mardon*

Water and Foam Carriers

Water carriers are often provided in areas where water supplies are short, and not only are they now more frequently seen, they are also getting bigger. Foam carriers provide a means of ensuring that an adequate supply of foam compound can be transported and made available within a brigade area when required.

Water carriers are by no means new to the fire service. In 1824 Glasgow Fire Brigade increased to eight the number of butts, or water carts, that conveyed water from the nearest fire plug to the engines, and they remained in use in Glasgow until the introduction of a water supply from Loch Katrine 35 years later. The dam lorry of World War 2 was a crude water carrier that was superseded after the war by purpose-built or adapted road tankers.

Right:

In 1967 Bedfordshire Fire Service prepared specifications for two new 1,200-gallon purpose-built water carriers, primarily for use on the M1 motorway and in smaller villages with limited water supplies. However, financial restrictions prevented the new vehicles from being ordered, and instead the brigade acquired two second-hand 2,000-gallon petrol tankers from Shell-Mex and BP Ltd. Sun Engineering built lockers for each to carry a 2,500-gallon inflatable dam and two portable pumps, fairing these into the lines of the tank to produce a smart design. *Charles Keevil*

Right:

Lincolnshire Fire Brigade was credited with having the only Foden fire vehicle in the country when it put this 1,500-gallon water carrier on the run in 1998. The Foden 2215 chassis and Whale Tankers tank were supplemented by additional bodywork by Warner of Lincoln. The appliance was based at Spalding retained fire station in the south of the county. *Geraint Roberts*

Transportable Water Unit

The transportable water unit was an AFS vehicle carrying three inflatable rafts and nine portable pumps designed to pump water from sources not readily available to land appliances. The rafts, each with three pumps, could be launched across soft ground or lowered by the vehicle's crane into water from where they could pump water ashore for fire-fighting purposes. The idea was conceived by the Home Office and development started in 1952 in conjunction with Kent Fire Brigade. The Bedford S prototype Bikini unit, as the appliance had become known, was allocated to the Emergency Training Centre in Reigate, Surrey, and in 1956 the first production unit on a Commer Q4 was handed over to Kent. Author's collection

Control Units

A control unit provides the officer in charge of a large fire or incident with a mobile office and communications centre on the fireground from which he can direct operations. It has always enabled him to maintain contact with brigade headquarters, but before and during World War 2 this might only have been by field telephone or connection into the public telephone network. Later the use of radio became standard, with development through fax to the sophisticated computer links available today. Equally essential is a means of communication with other officers and fire crews at the incident. Today this is undertaken with the aid of fireground radio systems, but before their availability messages would have been passed by the use of runners, or even motorcycle despatch riders if long distances were involved.

Once a control unit has been set up, appliances and crews arriving at the incident book in and a record is kept of resources and personnel in attendance. It is therefore necessary for the unit to be readily identifiable, and the customary marking was a red-and-white chequered panel on the body with a red flashing light at night. Later the panel became a chequered band round the bodywork supplemented by an internally illuminated red-and-white chequered dome on the roof. Various other means of identification have been used, including alternate red and white flashing lights, and nowadays a telescopic lighting mast with a flashing beacon and floodlighting is usually provided.

In the days before computers, maps and plans of the brigade area, water supplies and special risk premises were filed in cabinets, or kept on microfiche, and records and plans of an incident were kept on paper and drawn on blackboards. There is still a requirement for these traditional records, but much greater reliance is now placed on vehicle-mounted computers giving access to databases, geographic information systems and the like. Many control units are now equipped with closed-circuit television facilities and video recording, enabling an incident commander to view the scene from his mobile headquarters. Such is the level of equipment provided that in some brigades the control unit is built on a coach or similar-sized chassis with a body that can expand under hydraulic power to twice its normal width to provide adequate working space for the commander and his staff.

In 1994 Bedfordshire Fire & Rescue Service put a new incident command unit on the run. Based on the Iveco-Ford Eurocargo chassis with bodywork by G. C. Smith, the appliance was divided into the three usual areas, the front cab acting as a communications centre, a centre command area, and a rear conference area. The brigade's nominal roll boards incorporated personal issue tallies with individual barcodes, and crews reporting to an incident had their details scanned into a computer on board the command unit. This enabled not only a track to be kept of personnel on the scene but also the immediate availability of records of any previous exposure to radiation or chemicals, and their specialist skills. The appliance also provided video facilities for officers to monitor the fireground situation.

Right:
In 1949 Kent Fire Brigade commissioned a reconditioned 1931 Leyland Titan double-deck bus for use as a mobile control unit, canteen and exhibition unit. The lower deck control facility included a fire alarm recorder, telephone and switch for the call bells installed on both decks. This enabled the appliance to serve as a mobile fire station. It did just that at the county agricultural show when seven alarm boxes in the showground were connected to the recorder and a telephone was connected to brigade headquarters via a Post Office line. Radio was added later. The upper deck provided a canteen facility supplied with water from a 70-gallon tank. Six bunks in two tiers, two concealed wash-basins, a wardrobe and collapsible dining tables were fitted in an adjoining rest and mess room. An awning could be erected outside when the appliance was used for exhibitions and demonstrations. *John Meakins collection*

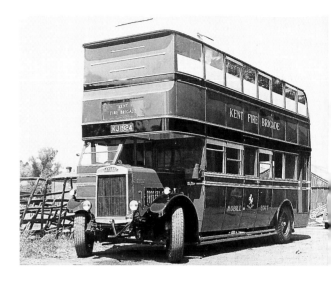

Left:
Surrey Fire Brigade found that this Morris chassis with a Wadham ambulance body met the requirements for a new control unit in 1964.
Author's collection

Right:
Avon Fire Brigade commissioned this Dodge S56/Spectra appliance in 1986. It is typical of the many mid-size control units in service at the time. *Roger C. Mardon*

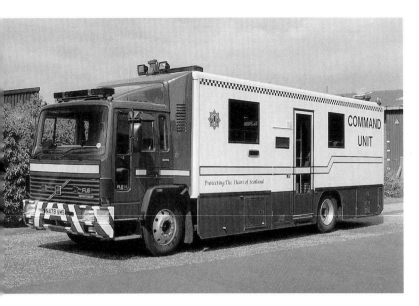

Left:
**In 1996 Central Scotland Fire
Brigade commissioned a new
command unit built by Heggies
Coachbuilders of Cupar on a
Volvo FL6.14 chassis. Teklite
460/8 lighting masts were
installed at the front and rear of
the vehicle, the rear one having a
dual red-and-white rotating
beacon for identification and
fitment for a video camera.
Internally, at the front the
appliance has the usual
communications area, and
behind is a command area
equipped with television and
video facilities. At the rear is a
BA main control compartment
with a work-desk and facilities
for cleaning BA masks.**
Gordon Baker

London Fire Brigade's command unit new in 1998 at a cost of some £150,000 was built on a Volvo B10 city bus chassis, with a mid-underfloor 10-litre engine and ZF automatic transmission, by Spectra Coachbuilding Ltd of Westbury. The 10.5m-long appliance is designed for use at incidents of eight pumps or more, providing a forward eight-seat planning and briefing area with computer and closed-circuit television monitor screens. The rear compartment contains four computer work-stations enabling operators to overlay resources assigned to an incident on an Ordnance Survey base and to access building plans and hydrant information. The two compartments are separately air-conditioned and telescopic masts provide radio aerial and CCTV camera facilities.

Requirements vary from brigade to brigade, and at the opposite end of the spectrum are Devon's four Mercedes-Benz Vito CD1110 vans fitted out as incident command vehicles in 2000 with VHF and UHF radio, cellular phone and fax, and a vehicle-mounted data system.

Canteen Vans

Fire-fighters at protracted incidents need hot drinks and food to keep them going and to offset the effects of the wet and often cold working conditions. As they are unable to leave the fireground, it has long been the practice for refreshments to be taken to the scene and provided from a suitable vehicle.

Rickmansworth's 1892 hose van and tender included a 'good canteen cupboard' below the driver's seat with access from doors on each side of the machine. The first reference to a canteen van in the reports of the Metropolitan Fire Brigade is in 1898, and certainly by 1922 the body of a horse-drawn canteen had been transferred to the chassis of a 1908 Commer hose tender.

In 1940 the Home Office recognised the need to keep firemen adequately refreshed at major incidents and sanctioned the acquisition of canteen vans by fire brigades, so long as it could be demonstrated that one could not be obtained from some charitable source. Strangely, the need was calculated by reference to the population of the brigade area rather than the needs of the fire-fighters, so that one canteen van for every 100,000 inhabitants outside the London region, or 75,000 in London, was approved provided it did not cost more than £130. Otherwise arrangements had to be made with a local pie-shop!

Canada was a prolific donor of canteen vans during World War 2, and such donations to the fire service from any source were always welcome. The presentation of two such vehicles by the Durham Branch of the British Legion drew a number of civic dignitaries and senior fire officers. 'I have seen the great value to hard-pressed firemen of a little sustenance at the right time,' said Chief Regional Fire Officer T. Breaks, whose name alone must have qualified him to comment with authority on the subject.

In the days of the NFS, and the postwar AFS, the provision of up to 200 hot meals at a time was within the purview of a bigger vehicle known as a mobile kitchen. Solid-fuel cookers were provided to maintain independence from oil or gas supplies, and coal would be carried on-board or in a trailer. A tank of water and a water heater were available, along with storage space for food, utensils and crockery. There was no standard vehicle in the war, but the AFS mobile kitchen was based on the Bedford S 5/7-tonner.

Above:
In 1935 London's new forward-control Dennis Ace canteen van was exhibited at Olympia. Serving hatches were made at the nearside for firemen and at the rear for officers, while the inside provided fittings for the accommodation of crockery,

foodstuffs and other items necessary to supply refreshment to fire crews at prolonged incidents. Two 10-gallon and two 5-gallon boilers were installed, with centrally fed Primus burners, a 40-gallon tank of drinking water, sink and oven to take 100 meat pies. *Surrey History Service/Dennis Bros*

Such vehicles are no longer used and the modern canteen van — or emergency catering unit as one brigade has chosen to call it — equipped with deep-freeze storage, Calor gas and microwave cookers, is able to serve the needs of crews at lengthy incidents. Not every brigade operates such a specialised vehicle, relying instead on the present-day equivalent of the local pie-shop, the fast-food outlet. Pumping appliances often now carry hot drinks facilities and limited pre-packed refreshments, reducing the need for canteen vans.

Demountable Bodies

Special appliances in fire service use can experience substantial periods of inactivity, and it was not necessarily cost-effective for coachbuilt vehicles to be constructed for the fulfilment of a single role. Interchangeable bodies or pods, which could be internally designed to requirements, provided an economic solution, and the range of units soon expanded to encompass most special uses. With any demountable pod system, different units are interchangeable on the same chassis, allowing one vehicle to be used for a variety of jobs. The demountable system does not create different categories of fire appliance, but merely an alternative way of providing them.

However, by 2000, after 25 years of service, Oxfordshire decided to discontinue the use of demountable units. Their use over that period indicated that control, incident support and environmental protection functions were in greatest demand, while bulk foam and other support equipment was rarely used. Pods were also considered vulnerable to high winds. As a result, the brigade reverted to self-propelled appliances, and the first three, for the high-demand roles, were based on the Mercedes-Benz Vario van.

Trailers

Trailers and semi-trailers are another way of using one prime mover for a number of different appliances. An articulated vehicle can, of course, provide a larger appliance than a rigid chassis, but there are more small trailers like the Indespension Tow-a-Van in service than there are articulated units. Caravans found favour with Cleveland Fire Brigade, for example, for use in a number of special roles such as decontamination and controls units.

In 1989 Tyne & Wear Fire Brigade bought three Tow-a-Van four-wheel trailers for the transport of specialist equipment. A structural collapse unit was equipped with a range of Acrow props, shoring timber, digging and cutting tools. A decontamination unit carried protective clothing, and equipment for dealing with chemical and radiation incidents, while the third unit was a BA servicing trailer. By 1990 Torton Bodies of Telford had built trailers for Grampian, Hertfordshire, Northampton and Lothian & Borders fire brigades.

Above:

In May 1966 Liverpool Fire Brigade acquired five Hands 8/10-ton ex-brewery semi-trailers, which were adapted in brigade workshops to fire service use. No 1 carried 900 gallons of foam compound in drums plus foam-making equipment, including one No 20 and three No 10 mechanical foam generators complete with hose and branches, and six No 10X foam-making branches. No 2 carried ship fire-fighting equipment and was originally very much like a low-loader with side rails, but was later rebuilt with side lockers. This responded as a first attendance to off-shore fires, and equipment was transferred to a tug or salvage vessel. Four light portable pumps, suction hose, loading ramps and slings were carried, with breathing apparatus, delivery hose, branches, lines, lights and life-jackets contained in five 18cu ft packing cases. No 3 (illustrated) was a heavy water trailer to carry large quantities of hose and other equipment that might be needed at large fires, including 100 lengths of 2¾in delivery hose, suction hose, hydrant equipment, 13 ground monitors, branches and breechings. No 4 comprised a canteen unit in the front half, equipped with a Calor gas cooker, a 100-gallon fresh water tank and heater, sink and cupboards. A dining compartment, which doubled as a mobile conference room, was in the rear. The fifth trailer was intended as a heavy rescue unit with cutting and spreading gear, heavy jacks, scaffolding and other tools, but was still in the workshops as a reserve in 1970. Two new 12-ton Austin tractor units were acquired in August 1966 and November 1967 to complete the articulated fleet.

Author's collection

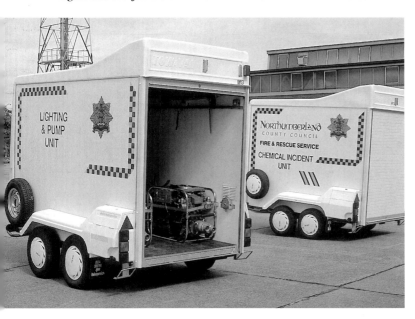

Left:

Northumberland's Tow-a-Van trailers of the 1990s included lighting and pump units and a chemical incident unit.

Roger C. Mardon

Other Miscellaneous Appliances

Rapid Intervention and All-wheel Drive

The armed forces had made use of cross-country vehicles for some time, and the coming of a light all-wheel-drive fire-fighting vehicle with similar capabilities was a boon. Several such appliances fulfilling much the same role as any other, but on a smaller scale and with greater manoeuvrability, have already been noted.

The four-wheel-drive Land Rover was introduced in 1948 as Britain's answer to the American Willys Jeep, and the following year Derbyshire Fire Service acquired four for use as personnel carriers. None carried a regular complement of fire-fighting gear, but it was envisaged that men and equipment would be transported over the county's rough ground to tackle moorland fires. Additionally the vehicles could be used to tow trailer pumps as necessary, and when not in use for fire-fighting they would provide ideal transport for personnel engaged on inspections.

One of the earliest Land Rovers to be equipped as a fire engine appeared at the 1952 Commercial Motor Show, and the following year the Ministry of Works acquired one with a Pegson 200gpm pump, 40-gallon first-aid tank and 120ft hose-reel for the Windsor Castle brigade. This appliance also carried a 35ft alloy extension ladder.

The Austin Champ four-wheel-drive vehicle was developed in 1952 for military use, and in 1956 Cornwall Fire Brigade acquired a Champ chassis and in the brigade workshops turned it into a light four-wheel-drive pump. The 10cwt 4x4 Austin Gypsy was introduced in 1958 and Cornwall will be better remembered for its development of this vehicle for fire-fighting use. The Coventry Climax ACP pump was also available in 1958 for front mounting on appliances and therefore tended to be fitted to light vehicles, such as the Austin Gypsy for which it was standard equipment. It was a single-stage centrifugal pump rated at 500gpm when operating at 100psi.

First-strike appliances, or rapid intervention vehicles, were developed primarily to provide some form of attack pending the arrival of heavier equipment and conventional appliances in areas of limited access, such as shopping precincts and multi-storey car parks, and to overcome delays caused by traffic congestion. They needed to be small and manoeuvrable, and were consequently built on Land Rover and Ford Transit-type chassis. Some may have been slower than their full-size counterparts, possibly because of a tendency to carry more equipment than was necessary, and crew accommodation was sometimes considered inadequate. It is probably fair to say that in urban areas first-strike appliances had some success in the 1960s and '70s, but were a compromise solution that did not last.

In works fire brigades, the first-strike appliance was usually designed to suit its particular location and application and would often be the only fire engine maintained on site. By the early 1980s the number of industrial brigades was noticeably on the decline and consequently the number and variety of works appliances has reduced.

The first Mercedes-Benz Unimog to see service was on the run as a dry powder crash tender at Manchester Airport in 1958.

The Steyr Puch Haflinger, a four-wheel predecessor of the now familiar 6x6 Pinzgauer, made its first appearance in the UK in 1964. Four years later, with bodywork by A. Schofield & Sons Ltd of Bury, and a 50-gallon tank, portable pump and hose-reel, one went into service with West Riding Fire Brigade. It was allocated to Sedbergh (then in Yorkshire, but within the new county of Cumbria from 1974), where it remained until 1981. London Fire Brigade undertook evaluation trials of a Haflinger 4x4 as a first-strike appliance in 1972. After familiarisation by use as a staff car in each division, the machine provided a supplementary response to the first attendance on fire calls, but its use was not developed after the trials had ended.

In 1988 Telehoist Ltd of Cheltenham appeared on the scene with fire and rescue vehicles built on the General Motors K30 four-wheel-drive chassis. In contrast to the steel cab, the bodywork was constructed of lightweight glass-reinforced plastic. The fire tender offered a five-man double crew cab and a pump capacity of up to 2,800l/min. A 900-litre water tank

Right:
Perhaps the most spectacular rapid intervention vehicle has been the 150mph Jaguar XJ12, adapted by Chubb to carry a 90-litre pre-mix foam tank in place of the rear seat, for protecting Thrust 2 during the World Land Speed Record attempt in Utah in 1981. The 1988 Jaguar 4-litre XJR pictured was provided by the car company for use at Donington Park race circuit, and was fitted with a Chubb 25-gallon AFFF installation. *Colin Dunford*

In 1964 the City of Coventry Fire Brigade acquired a Land Rover primarily for dealing with fires in multi-storey car parks inaccessible to full-size appliances. The Carmichael-built appliance incorporated a rear-mounted 500gpm Coventry Climax pump with a 40-gallon first-aid tank and 120ft hose-reel. A front-mounted power winch was also installed. *Carmichael & Sons (Worcester)*

Left:
This Ford Transit light pump was built by HCB-Angus in 1982 for the United Kingdom Atomic Energy Authority. The company had developed a power take-off that drove up through the floor behind the gearbox and connected to the Godiva FWPS pump via a drive-shaft. A 90-gallon water tank was installed and a 30ft alloy ladder was carried on the roof. *HCB-Angus Archives*

Left:
In 1991 Somerset Fire Brigade acquired a Steyr Puch Pinzgauer 718K 6x6 appliance to replace a Range Rover 6x4 at Dulverton. This was so successful that in 1994 another was acquired for Cheddar. Powered by a Volvo 2.4-litre diesel engine and engineered by Saxon Sanbec, the machine came with a demountable Godiva GP1600 pump and 500-litre tank. The cab accommodated a crew of five, and a 10.5m triple extension ladder was carried. The price of £83,000 included a driver training programme. *Roger C. Mardon*

was standard, with a 90-litre foam tank available as an option. The rescue tender provided a two-man cab, 110V electrical system and lighting mast. Heavy equipment was stowed on the centre-line of the chassis with access from the rear. Buckinghamshire Fire & Rescue Service ordered one of the first pumping appliances, while Gloucestershire had both pumping and rescue vehicles.

The Austrian-built Steyr Puch Pinzgauer Turbo D 4x4 or 6x6 appeared in 1989, and later in the year the first went to the Isle of Man Fire Brigade. This was a six-wheel-drive model with a 2,383cc six-cylinder diesel engine and fire engineering by Saxon. A 200-gallon water tank, 300gpm demountable pump and 10.5m ladder were provided, and a five-man crew could be carried.

Right:

At the Fire '95 exhibition, the Highlands & Islands Fire Brigade exhibited an LDV 400 van converted by brigade workshops into a light pump for use by volunteer units of the brigade. It was fitted with a demountable pump giving 1,000l/min at 10 bar and a 450-litre water tank. BA and rescue equipment was carried. *Geraint Roberts*

All-terrain Vehicles

In 1971 Surrey Fire Brigade acquired an Argocat, an amphibious plastic-bodied vehicle about 10ft long with eight independently driven wheels powered by a two-stroke 436cc engine. Balloon tyres inflated to only 2psi gave outstanding grip, and an outboard motor could be fitted enabling the vehicle to cope with 1-in-1 gradients, and cross swamp, marsh, liquid mud, sand and soft snow. Six men and a portable pump could be carried on this machine. Since then an assortment of all-terrain vehicles has been used by various brigades.

Right:

Scot-Track was established in 1988 and builds 8x8 all-terrain vehicles at Nairn. The wheels are designed so that tracks can be fitted over the tyres, and a range of options is available, including a crane with stabilising legs and a trailer, as seen here. West Yorkshire Fire Service put this HillCat 1700 on the run in 1998 and it is allocated to Huddersfield. It is transported in a curtain-side demountable pod. *West Yorkshire Fire Service*

Robots

Left:

In 1998 J. C. Bamford Excavators Ltd (JCB) developed a remote-controlled skid steer loader in conjunction with West Yorkshire Fire & Rescue Service. It entered service and is based at Dewsbury fire station, from where it is transported to incidents on a carrying vehicle. The machine is equipped with cameras capable of transmitting pictures to a forward control so that dangerous situations can be assessed from a distance without putting the lives of fire-fighters at risk. As originally developed the vehicle was equipped with an IFEX 3000 Impulse fire-extinguishing cannon, which discharged tiny — 2 micron — water droplets at up to 448mph, as an alternative to the grapple attachment. The grapple is used for remotely moving chemical drums and debris, for example, again to reduce the personal risk to fire-fighters. *West Yorkshire Fire Service*

6. WATER, RAIL AND AIR

Fireboats

Floating fire engines, or fire floats, were introduced not only to attack fires on ships and boats but also to fight fires in waterfront properties. The early vessels had no inbuilt means of propulsion and were towed to the fire by tugs. Quite when the first fire floats were brought into use on the River Thames, or elsewhere, is unknown, but in 1760 manual pumps were installed on barges to be rowed to riverside fires, and Shand Mason is known to have repaired one in 1793 and another in 1803. In 1812 and 1848 Tilley built fire floats for use in the London Docks. In 1837 the London Fire Engine Establishment put into operation a 54ft fire float provided with three double-cylinder pumps worked by levers running across the width of the boat and extending to 18ft in length when unfolded. The craft was propelled by oars and had a regular complement of 90 men. Tilley built a fire float for Sunderland and South Shields in 1841, and received orders the following year from the Admiralty for seven more to serve in the Royal Dockyards.

The first steam fireboat on the Thames was a tug of the West India Docks equipped for fire-fighting in 1850 by the installation of one of Downton's pumps, driven by the tug's propulsion engine. This threw 600 gallons of water a minute 20ft higher than the tallest warehouse in the docks.

In 1851 Braidwood and the London Fire Engine Establishment finally accepted the idea of steam power, and at his instigation Shand Mason arranged a Penn & Son donkey engine as a fire pump, with good results. At that time the London Fire Engine Establishment had two fire floats, one at Southwark Bridge with two 9in single-acting pumps requiring 80 men to work them, and a 60-man manual at Rotherhithe. The following year, Shand Mason converted the Southwark vessel to steam power by the installation of double-acting steam cylinders to drive the pumps. Water was also pumped to the stern and discharged under pressure so that the craft could be propelled by the thrust of the water jets. However, it could not be properly steered and propulsion by oars was reverted to. In 1854 it was decided to build a purpose-designed steam float. This was an iron vessel 130ft in length fitted with two Shand Mason steam engines, one each side, intended to power both the craft and the fire pumps. Propulsion was through the reaction of two 12in propelling water jets from an Appold centrifugal pump, but the vessel managed to attain only 8mph in tests and struggled to make any speed against the tidal flow of the Thames.

The two 10in reciprocating fire pumps, however, were capable of discharging over 1,900gpm through four 1¼in jets, and this fire float is reputed to have worked for nearly 400 hours at the Tooley Street fire in 1861. After the formation of the Metropolitan Fire Brigade in 1866, Captain Eyre Massey Shaw reverted to raft-mounted steam pumps.

The internal combustion engine became the normal power source, but otherwise fire floats such as these were built until 1940 with little outward change. Most were one-off designs; Bristol's *Pyronaut* was completed in 1934 and London's *Massey Shaw* the year after. Also in 1935 Merryweather supplied the fire pumps and foam sets for the Manchester Ship Canal's new tug fire float on which, interestingly, the two 500gpm 'Fire Suds' foam sets took water from a steam pump and air from a steam-actuated rotary compressor.

The first wartime fire boats were wooden vessels 40ft 6in long designed for use on the Thames and other tidal rivers. Powered by a 30hp V8 engine, they could achieve 10 knots. Two independently driven 1,100gpm pumps of the type fitted to land-based extra heavy units were installed, each with its six delivery outlets on deck. A monitor could be supplied by connection to eight of those outlets. From this 'Thames' class of boat the 52ft 'estuarial' type was developed to fight fires on ships entering the river as well as to operate in the upper reaches of a river where they might have to negotiate locks and low bridges. These were driven by two six-cylinder 70-75hp engines and provided with four 700gpm heavy pumping units. 'Canal' type boats 45ft in length powered by two four-cylinder 12hp engines and fitted with one 1,100gpm pump, but no monitor, were built with the ability to negotiate 7ft locks. They manoeuvred equally well in either direction and could be controlled from either end.

In keeping with the wartime spirit of improvisation, fire boats large and small were converted from a variety of vessels from trawlers to cabin cruisers. Until World War 2 Britain had no sea-going fire boats, but they became necessary for fighting ship fires at sea and were also used in protecting their home ports.

It was the terminology of the National Fire Service in 1941 that brought the designation 'fireboat' into common usage, in preference to fire float, which suggested a lack of self-propulsion.

Many wartime fireboats continued to give service after the war, but new craft were also commissioned in postwar years. For example, in 1956 it was announced that Glasgow Fire Service was to have a new all-steel

R. S. Hayes (Pembroke Dock) Ltd was given an order by the BP Tanker Co in 1959 for a fireboat to be used in Swansea Harbour. A twin-hulled vessel with a 25-ton-capacity foam tank and fire-fighting tower with nine Merryweather foam/water monitors was specified. Output was to be 3,100gpm of water and 12,500gpm of foam. The result was the *BP Firemaster*, which was put into service in 1960. Two 140bhp Dorman diesel-powered Harbormaster propulsion units enabled the vessel to move ahead or astern under full power, as well as crab-wise. Large pneumatic wheels were fitted at each end of the twin hulls enabling the craft to move along the side of a burning vessel. *Author's collection*

fireboat at a cost of £80,000 to replace a wooden vessel built in 1940. In 1958 an order was placed with Hugh MacLean & Sons for a twin-screw craft, with pumping equipment able to draw water from the Clyde at the rate of 32 tons a minute, to be supplied by Merryweather. The following year the 68ft *St Mungo* with a working speed of 9 knots was put into service.

SARO (Anglesey) Ltd built London's new fireboat, *Firebrace*, which was handed over in 1961, and Southampton had its new 65ft *Fireboat 39* in 1963, built by John I. Thornycroft & Co with fire-fighting equipment by Merryweather & Sons.

Said to be the first fire-fighting vessel in the world fitted with a hydraulic platform was the tug *Coleraine*, built in 1969 by Appledore Shipbuilders Ltd for service in Belfast Harbour. The fully extended Simon Snorkel platform was 75ft above water level and equipped with a Merryweather monitor able to discharge 1,600gpm of water or 5,000gpm of foam. Two additional deck monitors, a 3,000gpm Merryweather MM2-10 pump powered by a Rolls-Royce 436bhp marine diesel, and a 4,500-gallon foam tank were installed.

Certainly new boats continued to be built, but soon the

cost of maintaining them would assume an overwhelming priority. In 1963 Liverpool Fire Brigade estimated that it was costing £1,000 a week to maintain its fireboat and withdrew the *William Gregson* from service. Instead the brigade arranged for portable pumps and crews to be put aboard Mersey tugs and salvage vessels when a water-borne attack was required. Arrangements with tug companies for the use of their vessels equipped for fire-fighting were destined to become the norm for most brigades with waterfront risks.

In 1976 Humberside Fire Brigade ordered a £62,000 fireboat from Nordic Star Marine of North Yorkshire to replace the *Clara Stark*, which had served on the Humber since 1939. The new 50ft shallow-draught vessel was powered by twin Dowty water jets driven by two marine diesel engines. A Ford-engined 1,000gpm Godiva pump and two auxiliary 350gpm fire pumps were installed, together with a water/foam monitor and a 7,000cu ft per minute foam generator. *Fire Jet 1* would normally have a crew of four. In 1980 the new boat was up for sale as a result of local government spending cuts, and arrangements had been made for fire cover to be provided by a local tug company.

The last 20 years have seen a rapid increase in fire brigade use of fast, inflatable and rigid inflatable boats on estuarial and inland waters. Kent Fire Brigade commissioned two of these 7.9m Delta rigid inflatables in 1998, and they can be launched from trailers at many sites throughout the county. Powered by two Honda 90hp engines, these craft have a top speed of 35 knots and are equipped for 24-hour operation with satellite navigation, echo-sounder and radio links with brigade and port control. The total crew of six has available a 3.8kW generator and floodlights, two Grindex 550l/min submersible pumps, 90 litres of AFFF and four breathing apparatus sets. A six-person inflatable life raft is also provided. *Kent Fire Brigade*

When London Fire Brigade took delivery of two Alnmaritec boats in 1999, the crew of three would be responsible for transporting land-based personnel to an incident and, in a change of practice, would not themselves be responsible for the fire-fighting response.

Railway Fire-fighting

With timber carriages illuminated by gas and freight wagons hauled by coal-fired locomotives, it is hardly surprising that the early railway companies were mindful of the risk of fire. It was a threat not only to the trains but also to railside crops and grass, which were frequently set alight by sparks from the engine. The advent of spark arresters for steam locomotives, and indeed fire engines, reduced the problem but did not eliminate it. Fire trains were devised to provide protection for the property at risk, often in areas beyond the reach of any fire brigade.

In 1842 Merryweather constructed a manual fire engine for use on the rails of the London & Birmingham Railway. At the time it was the largest manual engine ever built for land service, with a 450-gallon cistern and two 9in pumps having a 10in stroke. It could carry 400ft of hose on a hose-reel at the front and, when operated by 42 men, a good jet of water in excess of 100ft high was produced. In 1861 Shand Mason built three double-cylinder horizontal steam fire engines to run on rails for the London & North Western Railway.

Later fire trains comprised a tank locomotive to haul a wagon with a steam pump, hose and equipment, water being supplied from the engine itself. Developments led to red-painted trains hauling a 3,000-gallon water tank wagon, a coach with trailer pump, a mile of hose, and equipment, then a coach providing crew accommodation and amenities. Steam pressure was kept up in the engine, at 40lb below working pressure, ready for a quick response.

By 1930 the railway companies were replacing wooden carriages with steel ones and changing over from gas-lighting to electricity in an attempt to minimise the risk of fire, especially following a train crash. Nevertheless they had to remain prepared for the worst, and at Crewe, Derby and Horwich the London Midland & Scottish Railway maintained a locomotive tender holding about 5,000 gallons of water and a brake-van in readiness for emergencies. The brake-van carried a portable pump, and in 1930 the company took delivery of its third Dennis 250gpm pump for such application.

Extra fire trains were put in service during World War 2, and the Southern Railway established a number comprising large water tanks drawn by steam locomotives fitted with Merryweather 'Hatfield' pumps.

In 1952 the three ex-LMS fire trains were still employed by British Railways, but they were expensive to maintain and had outlived their usefulness. Local authority fire brigades were considered able to cope with incidents, and the use of fire trains on the public network was discontinued.

The Channel Tunnel poses special risks for fire-fighting and rescue, and purpose-designed vehicles were built to operate in the service tunnel between the two running tunnels. While these vehicles are not themselves trains, they are designed to operate in the environment of a rail tunnel and it is convenient to look at them here. The Service Tunnel Transport System (STTS) is based on 1993 Mercedes-powered units with a driving cab at each end and interchangeable body units for fire, ambulance, police and maintenance purposes. Four STTS fire appliances, with bodies by John Dennis Coachbuilders, are stationed at each of the English and French terminals, together with two ambulance vehicles and one police communications unit. Two First Line of Response (FLOR) appliances (as illustrated), each with a crew of four, carry fire-fighting, breathing apparatus, rescue and first-aid equipment. Two Second Line of Response (SLOR) vehicles, with accommodation and BA sets for 12 crew, transport personnel and equipment from supporting Kent and Calais appliances, with one (STTS 3) operating as a forward control unit.

FLOR appliances can attend any incident

Right:
A Mercedes STTS vehicle with fire-fighting body at the English terminal of the Channel Tunnel. These vehicles are registered with both the British and the French licensing authorities and therefore display a registration mark from each country. *Roger C. Mardon*

throughout the full length of the tunnel, but SLOR vehicles cannot cross the English/French border until requested after an incident has been made bi-national. Six pumps initially respond from Kent Fire Brigade to tunnel incidents on the English side, with a seventh attending at the Incident Command Centre on the terminal site. Crews and whatever equipment may be needed are transferred to the SLOR appliances for transport through the service tunnel to the incident, and STTS 4 is available to shuttle personnel and equipment to and from the terminal Emergency Response Centre. Alternatively, road appliances and their crews can be ferried to an incident through the running tunnel on a goods vehicle shuttle train.

Aircraft and Helicopters

Fixed-wing aircraft have played a part in the fire protection of North American forests since 1919, and a year earlier San Diego was using an aero first-aid tender, based on a seaplane, primarily to protect its waterfront properties. In the UK the deployment of aircraft has been limited. In 1985 North Yorkshire Fire Brigade reported the successful use of a light aircraft, made available by the Cleveland Flying School, for reconnaissance of a moorland fire extending over a 3-mile front. The overall extent of the blaze could be readily assessed, enabling the brigade to identify and protect dwellings and farm buildings in its path. Two years later a MacAvia-BAe 748 was modified at the Cranfield College of Aeronautics to carry a pannier tank containing 2,000 US gallons for fire-fighting. The pannier could be removed or refitted at short notice and incorporated eight dump doors and a system of variable dropping rates.

The rising popularity of the use of turntable ladders and self-supporting escapes as water towers had demonstrated the advantages of fighting fire from above the building involved. Chief Officer A. Girdwood of Paisley was therefore prompted to observe in 1930 that developments in aviation meant that the aeroplane might become a fire-fighting unit, especially if a machine with hovering power was devised.

While fire brigades did not then aspire to having their own helicopters, by the 1960s it was not unusual for them to have made arrangements with operators for aircraft to be made available when needed. In recognition of this, in 1968 Pyrene introduced the Fire-Chopper, a container unit with 200-300lb of dry powder, BCF or 30 gallons of light water, for suspended transport below a helicopter to the scene of an incident. In operation the unit would be landed and one man would run out the hose while another operated the release valve for the extinguishing agent.

A study group established by the Joint Committee on Fire Brigade Organisation in 1976 produced a report, published in 1980, concluding that helicopters and light aircraft could assist fire brigade operations, but were not fundamental to their success. In spite of the report's generally negative comments about the cost of airborne hose laying and the effectiveness of water bombing, in Scotland the Northern Fire Brigade workshops constructed a hose-reel capable of holding 2,000ft of delivery hose, which could be slung below a helicopter and laid out in 2½ minutes. Trials were so successful that by 1980 five hose-laying devices had been constructed and strategically sited throughout the 12,000sq miles covered by the brigade. The experiments were continued by the brigade's successor, Highlands & Islands Fire Brigade, and the use of aircraft supplied by PLM Helicopters for reconnaissance and water bombing of forest fires, from a water bucket suspended below the helicopter, became established.

A helicopter from RAF Coltishall was ordered to a serious fire at an Ipswich power station in 1982 as an insurance against the possible failure of ladder and BA rescue teams which had been sent to reach a man trapped on the roof. In the event both rescue teams reached the man at the same time and the helicopter was not needed, but this was at a time when the use of helicopters at fires in buildings was virtually unheard of in this country.

Fire service thinking on the use of helicopters up to that time had been centred on the securing of arrangements for the availability of an aircraft on an ad hoc basis. By the mid-1990s there was recognition in some quarters that the helicopter as a permanent asset might bring to the fire service the advantages already bestowed upon the police and ambulance services. In 1994 Strathclyde Fire Brigade undertook a week-long evaluation of the Eurocopter BK117 helicopter, supplied by McAlpine Helicopters Ltd of Kidlington. The aircraft proved to be an invaluable aid to the mobilisation of personnel in a brigade area with a

coastline longer than that of France and remote islands 80 minutes away from conventional reinforcements. The 150mph helicopter could carry nine fire-fighters, or four with half a ton of equipment.

The following year West Sussex Fire Brigade undertook an operational assessment of a similar aircraft and concluded that the brigade's service to the community would be greatly enhanced by the provision of a helicopter. This trial highlighted the value of strategically located centres where equipment and personnel could be picked up for an even more rapid response, with the potential for reducing the deployment of land-based personnel and equipment. Needs were also identified for well-sited refuelling facilities to avoid a return to base during protracted incidents, and for incident commanders to recognise that support from the air was virtually instantaneous.

Following withdrawal of the search and rescue unit at RAF Bawdry, Mid & West Wales Fire Brigade undertook evaluation trials in 1996 with the Eurocopter BK117 and, as with other brigades, the value of the aircraft was confirmed. However, an initial purchase price of between £2.5 and £3.5 million, with annual running costs of £0.75 million, was likely to put acquisition beyond the reach of that, or any other,

brigade. However, this chapter in the progress of the British fire service is surely not closed yet.

Hovercraft

It was suggested in 1963 by the chief officer of the Isle of Wight Fire Brigade that the high speed and amphibious nature of air-cushion vehicles appeared to make them suitable for a fire-fighting role, especially in dock areas and over heath and marshy territory. Both the Westland SRN5 and the Hovermarine HM2 were the subject of design exercises, but, in spite of these ideas, fire-fighting from hovercraft appliances has not featured at all. They are, however, employed as rescue craft.

In 1984 Heathrow Airport, which lies close to reservoirs and sewage lagoons, commissioned a Tiger 4 hovercraft capable of 30 knots over water. The 16ft 9in craft was intended to carry two firemen and a 20-man life-raft to an incident and, if necessary, to run a shuttle service with further life-rafts. It was replaced in 1996 by a Hovertechnics Hoverguard 800 powered by a General Motors Crusader 160hp engine and carrying enough fuel for up to 5 hours of use. The craft was normally intended to be transported to an incident on a trailer, and carried two self-inflating 25-person life-rafts. Further rafts were available for collection at strategic points around the airport.

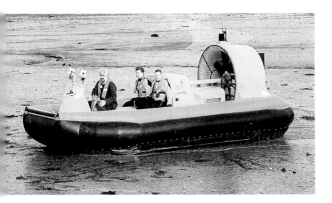

Left:
In 1999 Dundee Airport bought a Griffon 375 turbo-diesel hovercraft to provide rapid access to a line of inter-tidal land and water within the area of the airport fire service. The 30-knot craft, capable of carrying five people and powered by a Land Rover 300TDi diesel engine, was launched under its own power down a ramp that extended below the tidal mud level and was intended to give immediate help at an incident until further resources arrived. It carried a 30-man inflatable life-raft and was fitted with lights for night searches. Radio contact could be maintained with the airport fire service, HM Coastguard and the lifeboat service. *Griffon Hovercraft*